Geotechnologies and the Environment

Volume 14

The “Geotechnologies and the Environment” series is intended to provide specialists in the geotechnologies and academics who utilize these technologies, with an opportunity to share novel approaches, present interesting (sometimes counterintuitive) case studies, and most importantly to situate GIS, remote sensing, GPS, the internet, new technologies, and methodological advances in a real world context. In doing so, the books in the series will be inherently applied and reflect the rich variety of research performed by geographers and allied professionals.

Beyond the applied nature of many of the papers and individual contributions, the series interrogates the dynamic relationship between nature and society. For this reason, many contributors focus on human-environment interactions. The series are not limited to an interpretation of the environment as nature per se. Rather, the series “places” people and social forces in context and thus explore the many sociospatial environments humans construct for themselves as they settle the landscape. Consequently, contributions will use geotechnologies to examine both urban and rural landscapes.

More information about this series at http://www.springer.com/series/8088

Jay D. Gatrell • Ryan R. Jensen
Mark W. Patterson • Nancy Hoalst-Pullen
Editors

Urban Sustainability: Policy and Praxis

Springer

Editors
Jay D. Gatrell
School of Environmental Studies
Bellarmine University
Louisville, KY, USA

Ryan R. Jensen
Department of Geography
Brigham Young University
Provo, UT, USA

Mark W. Patterson
Department of Geography and Anthropology
Kennesaw State University
Kennesaw, GA, USA

Nancy Hoalst-Pullen
Department of Geography and Anthropology
Kennesaw State University
Kennesaw, GA, USA

Geotechnologies and the Environment
ISBN 978-3-319-79908-7 ISBN 978-3-319-26218-5 (eBook)
DOI 10.1007/978-3-319-26218-5

Printed on acid-free paper

This Springer imprint is published by Springer Nature
The registered company is Springer International Publishing AG Switzerland

Contents

Chapter 1
Urban Sustainability: Perspectives on Change

Jay D. Gatrell and Mark W. Patterson

1.1 About This Book

Sustainability resides at the nexus of philosophy, practice, politics, and technology. As such, the policies, behaviors, and discussions associated with "sustainability" are often contested. Despite the potential for conflict, nearly all stakeholders at all scales recognize that sustainability and sustainable practices necessarily extend from and/or interact across three domains: Social Systems, Ecological Systems, and Economic Systems. The systems, most often referred to as the three pillars (Environment, Economic, and Social), are sometimes complementary; but often are juxtaposed in a functionally perpendicular fashion. As a result, theory and practice are often side-ways and often result in partial or imperfect results.

This book assembles a collection of papers that presents alternate approaches or views to understanding urban environments and sustainable initiatives across all three domains at the local and regional scale. In nearly all chapters, remote sensing technologies and techniques from GIScience are used to investigate, visualize, and understand social, economic, or environmental dynamics associated with sustainability on the ground and in place. While every effort has been made in this collection to broaden notions of sustainability, the book is not exhaustive—nor is it intended to be. Likewise, the book doesn't set out to chart a history of urban sustainability. Rather, the novel spatial applications, cases, and policy discussions are intended to provide the readers with new ways of thinking about sustainability on the ground and in place.

J.D. Gatrell (✉)
School of Environmental Studies, Bellarmine University, Louisville, KY, USA
e-mail: jgatrell@bellarmine.edu

M.W. Patterson
Department of Geography and Anthropology, Kennesaw State University,
Kennesaw, GA 30144, USA
e-mail: mpatters@kennesaw.edu

J.D. Gatrell et al. (eds.), *Urban Sustainability: Policy and Praxis*,
Geotechnologies and the Environment 14, DOI 10.1007/978-3-319-26218-5_1

1.2 Chapters

This collection is comprised of 15 substantive chapters organized (or at least conceptualized) in three parts. Part 1—Environmental Change and Urban Morphology focuses on more traditional geotechnical applications of the natural and built environments within the context of broader sustainability initiatives and/or the implications of urban land use change on the overall sustainability of cities. Part 2—Economic Change, Industry, and Sustainable Local Alternatives examines industrial adaptations, innovations, and practices associated with sustainability. In the process, the contributions include novel examples associated with non-traditional industries such as brewing, urban agriculture, and metal recycling. The papers will deploy descriptive statistics, mapping, and other visualization techniques. Part 3—Socio-Political Change and Adaptation draws on concepts from political ecology, new social movements, and even cultural geography to understand observed change on the ground—as well as the broader implication of urban environmental change on the every lives of urban and peri-urban residents.

In Part 1, five chapters provide a solid foundation for understanding and assessing urban environmental systems. The contributions examine urban water management issues (Chap. 2 Lawrence), the emergence of new inter-disciplinary approaches towards assessing and managing complex eco-systems (Chap. 2 Portman and Elhanan) and the several real world examples rapidly changing urban systems in multiple contexts using a variety of tools (Lulla and Jensen, Jensen and Hardin, and McLean et al.). In concert, these five chapters underscore the importance of assessing urban environmental change over time and across space (Table 1.1).

The six chapters in Part 2 focus more squarely on economic systems and the many ways in which cities, their residents, and industries have responded to the realities of globalization, increased competition, and historical market failures. The papers examine multiple facets of dynamic local food systems (Chaps. 7, 9, and 11), an investigation of a sustainable practices in urban craft breweries (Chap. 10), and even novel adaptations of residents to economic decline that promote sustainability (Chap. 8). While the topics of the chapters may appear to be disparate, each of the authors interrogates the empirical realities of economic responses to change and efforts to enhance the long term sustainability (or viability) of economic systems and/or strategies to redefine local economics that enhance urban ecologies. In the process, the contributions provide the reader with new ways to visualize (i.e., map) social networks, think about urban agriculture, and explore the real world implications of historically invisible industries like scrap metal recycling.

The final chapters in the book—Part 3—focus on the politics and decision making processes surrounding urban environmental change and sustainability. Lascell (Chap. 12) details the social and political movements and sustainability initiatives that have emerged in New Orleans since Hurricane Katrina devastated large portions of the city—notably the poorest and most vulnerable communities. In Chap. 13, Fuller examines the full range of environmental justice issues facing poor

Table 1.1 Summary of substantive chapters in this book

	Author(s)	Subject
2	Lawrence	Urban stream management
3	Portman and Elhanan	Ecosystem services assessment
4	Lulla and Jensen	Hyperspectral applications
5	Jensen and Hardin	Urban forests in arid environments
6	McLean et al.	Greenspace and sustainability in Las Vegas
7	Reid	Local food systems, sustainability, and networks
8	Chohaney et al.	Sustainable economic development and urban "Mining"
9	Hagelman et al.	Urban gardens and sustainable agriculture
10	Patterson et al.	Sustainability and breweries
11	Knigge et al.	Urban food hubs and sustainable local food
12	Lascell	Sustainability in Post-Katrina New Orleans
13	Fuller	Mapping environmental justice
14	Gallagher and Lawrence	Unmanned systems
15	Metcalf et al.	Greening as a social movement
16	Clay and Albers	Planning sustainable built environments

and working class neighborhoods as they struggle to enhance their personal environments, local economic conditions and the overall sustainability of neighborhoods. Chapter 14 presents the case for the use of unmanned systems to assess urban change and broader sustainability initiatives across space. In the final two chapters, the politics of greening (Metcalf) and decision making (Clay and Albers) are considered within the context of broader issues of sustainability across all three domains.

Chapter 2
Urban Stream Management Using Spatial Approaches for Stream Clean-Up Data

Patrick Lawrence

Abstract In recent decades many local communities, supported by state, federal and/or international programs, have engaged in efforts to remove human debris from urban stream systems. Typically these clean-up events have involved volunteers who collect garbage and other waste from stream banks or from the water. The aims of these programs are to improve overall stream conditions – especially aesthetics – and remove harmful materials from stream environments including tires, plastics, metals and other potential items of concern such as paint cans, and rubber products. Although many of these efforts report basic data on items collected, with the reports and information often submitted to agencies or the focus of media reports, beyond basic types and numbers of information collected, the data and report do not typically contain any geospatial aspects such as locations, areas cleaned, collection of specific items tied to locations, or addressing possible sources for the debris. Since 1997, local groups within the Maumee Area of Concern in northwest Ohio, USA have been organizing an annual stream cleanup event in their communities that has evolved to over 1000 participants working at more than 60 sites covering 4 streams. This chapter examines the results for the Ten mile Creek/Ottawa River clean-up sites using detailed site specific data from 1995 to 2006 that includes items collected and recorded on data forms and then compiled by location and types of items and examined in reference to spatial aspects of management actions including considering potential sources and addressing local land use and human activities contributing the specific items collected at locations along the stream. Results include identifying the top ten items collected and examples of locations where items can be tied to adjacent land uses for purposes of identifying actions to address continued and persistent sources of debris and needed responses. The recommendations and proposals based on this study are intended to inform decision-makers not only at the local scale but to influence how stream clean-up data can be utilized and to improve reporting of this information. And with the aim of encouraging the collection of geospatial and location aspects as a means of furthering utilization of urban stream clean-up data to support and assist management actions to address aesthetic aspects of urban stream environmental improvements and rehabilitation efforts.

P. Lawrence (✉)
Department of Geography and Planning, University of Toledo, Toledo, OH, USA
e-mail: patrick.lawrence@utoledo.edu

J.D. Gatrell et al. (eds.), *Urban Sustainability: Policy and Praxis*,
Geotechnologies and the Environment 14, DOI 10.1007/978-3-319-26218-5_2

Keywords Urban • Water • Debris

2.1 Introduction

Among the many challenges facing urban streams is the growing issue of debris (or garbage) from various human activities and sources that end up in the waterway, along the banks, and within the riparian zone (which defines the edge of stream ecosystem). Common examples of this debris are paper and plastic products such as food containers and wrappers, construction materials, cigarette items, glass items, personal care products, household materials, clothing, and many other items discarded from residential, commercial or industrial sites typically found in urban settings (American Rivers n.d.). These items cause concern for their impact on visual or aesthetic appearances and as potential sources of water contaminants as they degrade in the natural environment (Allison et al. 1998; Burres 2009; Stickel et al. 2013). For example, the byproducts from the slow, but eventual, decay of rubber, plastic and other synthetic materials can be introduced into the stream aquatic environment. In recent decades concerted efforts have been undertaken in many urban communities to remove stream debris through directed local clean-up events organized by concerned groups and organizations (Riley 1998). These efforts mirror similar programs that focus on coastal and ocean debris.

In 2013, the Ocean Conservancy reported that the International Coastal Cleanup effort held worldwide resulted in 648,015 volunteers in 92 countries removing more than 12.3 million pounds of trash. The International Coastal Cleanup started in 1987 and has expanded to cover almost 13,000 miles of coastal and river/streams in more than over 80 countries worldwide (Ocean Conservancy 2014). In terms of cleanup of debris specifically from urban rivers and streams, several major programs have developed in many areas, including in the United States within California, Illinois, Wisconsin, Michigan, Ohio, Maryland, Pennsylvania, and New York State (Wang 2005; Missouri River Relief n.d.; Susquehanna River Basin Commission 1996). Internationally, examples include Australia, New Zealand, Israeli, Malaysia, and the United Kingdom.

Riley (1998) provides a comprehensive review and discussion of the challenges facing urban streams and rivers and provides science and community based solutions, that includes the need to remove unwanted debris.

> A *major portion of stream problems can be corrected by removing garbage, junk, and dumped waste from stream channels. Debris can deflect stream flows, causing significant bank erosion…may pollute the water… definitely destroys the aesthetic values of urban waterways*; *and it can back up flows, causing flooding*
>
> *Neighborhood stream cleanup projects are likely the most cost-effective flood damage reduction and water quality control projects a local organization can invest in* (*p. 328*).

Addressing debris from human activities has been the focus of numerous programs and policies within urban communities. The California Coastal Commission (n.d.) has produced a management plan to assist local municipalities in addressing trash and debris from storm water and urban runoff before it ends up in ditches and streams. Their recommendations include installing catch basin screens, netting, separators, litter or trash booms, and using anti-littering enforcement and education, along with organized community volunteer cleanups. The California Water Boards (2007) completed extensive assessment of trash found in streams within the San Francisco Bay region to identify potential sources and recommend management and regulatory solutions. The U.S EPA (2011) produced an informational flyer highlighting the issues and challenges with trash found in urban and coastal communities and how streams and rivers can be a significant source of garbage that can be harmful to aquatic ecosystems and the associated economic impacts from the subsequently polluted and degraded waterways. The Maryland Department of Environment (2010) completed an extensive study of daily loads of trash into the Anacostia River, pointing out the need for addressing sources.

For international examples, Golik and Gertner (1992) counted litter removed from six beaches located in Israel noting a density of 36 pieces per 5 m transect and that 70 % of the litter consisted of plastic with wood, metal, glass, and other materials making up the remainder. Their results suggested that proximity to a population center resulted in an increase to the amount of beach litter and that the presence of waves and storm events resulted in re-accumulation of litter on beaches frequently throughout the year. They also noted that unlike other coastal areas where ocean shipping and dumping were sources, within their study it was apparent that the recreational users of the beach were the main contributors as the litter consisted of items resulting from their activities (items included food and beverage materials, cosmetics, plastic bags, toys and garments).

It is important to note that in reference to this study, the focus on urban stream cleanup programs is on the sources and types of urban stream debris or garbage that is the result of human activities, and not the removal of woody debris or other natural materials that can also accumulate in urban streams. There is considerable debate and disagreement as to the need and the potential harmful impacts from the intentional removal of woody debris (such as tree falls) from stream and river systems. Although some programs and efforts are in place to remove such materials from streams due to their potential for backing up water flow resulting in upstream flooding or diverting stream flow energy into an adjacent stream bank creating potential for increased erosion, the practice of removing woody and other forms of natural debris from urban streams remains controversial. The New York State of Environmental Conservation (n.d.) has provided guidelines for the removal of woody debris and trash from rivers and streams where the debris could impact water flows by blocking bridge and culvert openings, diverting streams and causing bank erosion. They recommend that large woody debris (trees) found in the stream should be removed when it presents a risk to infrastructure, bridges or homes. Lassettre and Kondolf (2012) promote whether possible the conservation of woody debris in urban streams channels in order to maintain their important and well documented benefits as forms of natural habitat, sources of organic matter, and potential food sources. Readers interested in more details and discussion focusing on the specific issue of removal of woody debris from urban streams are referred to Larson et al.

(2001) or NSW Department of Primary Industries (2007), and research focused on the role of woody debris, for example Marcus et al. (2002).

Another view is taken by Nemeth and Keirsey (1999) who argue that organizing debris removal along stream channels, including woody debris, can serve little purpose if the streams remain stressed and devoid of life due to other human impacts; so why undertake the effort. The aim to make urban streams more "scenic" could be impacting the aquatic health of such natural systems. Nemeth and Keirsey (1999) extend their comments further to suggest that even the removal of human debris, such as wooden pallets, shopping carts, and tires could be seen as a means of creating aesthetic benefits that masks more serious environmental concerns associated with contaminated urban streams and related significant water quality concerns all too often present in these environments.

However, for the purposes of this study, the focus will be on how organized efforts to remove debris from human sources are undertaken by community organizations along urban streams. Of specific interest are whether data on items removed are collected and intended to be utilized to address source areas to implement voluntary or regulatory means to eliminate the sources so as to stop continued and persistent locations of stream debris? If such efforts are intended, what measures and methods are used to collect and organize the data and are any spatially based approaches used to identify stream sites where debris is found and collected, along with noting proximal source locations for this debris to determine appropriate follow-up management actions.

2.2 Stream Clean-Up Programs

The Ocean Conservancy prepares annual reports on International Coastal Cleanup (ICC) efforts worldwide, but does not provide an indication of the total number of clean-up programs or sites or distinguish between coastal and rivers/streams. However, the data do provide some insight as to the number of cleanup events and types of debris collected. During the 2013 ICC event, 648,015 volunteers covered 12,914 miles of shoreline and rivers/streams, removing 12.3 million tons of debris (Ocean Conservancy 2014). The most common items collected were: cigarettes, food wrappers, plastic bottles and caps, straws, plastic bags, glass bottles, and beverage cans. The 2014 report also includes case studies and highlights from several locations and examples of efforts undertaken to reduce the amount of debris found at some sites. Beyond summary data by country (and by state within the U.S), there is no spatial context to the reporting of collections or responses to address sources.

Information on specific individual stream cleanup activities also appears to be limited, with most materials consisting of promotional items intended to inform and recruit volunteers and instructions/guidelines for conducting a cleanup event. The Western Michigan Environmental Action Council (n.d.) provides detailed instructions on how to undertake a stream cleanup, focusing on organizing volunteers, insurance and safety issues, gaining permission to access properties, materials and supplies, and planning the event. Water Action Volunteers (2001) provides a simple ten step guide to conducting a stream and river cleanup, including list of materials and promotion of the event as well as the results and outcomes. In June 2011 the Friends of the Los Angeles River and Friends of the Chicago River (2011) produced a media release highlighting their friendly challenge to gather the most volunteers for their respective cleanup events. Since 1992 Friends of the Chicago River have collected thousands of pounds of garbage at 65 sites, while starting in 1989 the Friends of the Los Angeles River volunteers worked 15 sites, averaging 15 tons each year.

These types of summaries for urban stream cleanup activities appear to be the most the common form of reporting, with little information on specific sites or identification of sources areas that could be targeted for follow-up actions. In some cases urban stream cleanups are often undertaken in the context of larger regional watershed efforts as shown by such work undertaken within the Susquehanna River in Pennsylvania, where a local river basin compact was signed that dictated any dumping or littering upon the waters of the river or its tributaries was prohibited and was to be enforced by law enforcement officials (Susquehanna River Basin Commission 1996). There is very little evidence, reporting or documentation that urban stream cleanup efforts include methods or procedures – beyond counting number of volunteers, number of sites, miles cleaned, bags filled or weight of debris collected – to accurately document site specific results in terms of what is collected, where it is collected, and the identification of potential sources for the debris.

The focus of this study was to conduct a review of site level data from a multiple year period of cleanup activities within the Tenmile/Ottawa River watershed on the Maumee Area of Concern, located in northwestern Ohio, US. An annual stream cleanup event has been undertaken there since 1997, with detailed site information collected on types and total numbers of items collected by teams of volunteers. To date beyond summary reports by watershed, no detailed examination and analysis has been completed of the data from this cleanup activity. In addition, by working at the scale of the collection site it will be possible to examine the spatial context for items collected and consider proximal locations as sources for the debris collected. Such a study will provide opportunities to discuss and recommend how spatial aspects of urban stream cleanup activities should be examined and the benefits provided by such approaches in terms of future planning of subsequent cleanup events and addressing potential continued dumping or other site specific issues associated with the persistence of debris located at cleanup locations.

2.3 Maumee Area of Concern Clean Your Streams Event

Partners for Clean Streams (PCS) is the regional non-profit watershed organization in Northwest Ohio. PCS aims to strive for abundant open space and a high quality natural environment; adequate floodwater storage capacities and flourishing wildlife. Through various programs and activities, PCS encourages stakeholders to take local ownership of their aquatic resources, striving for local rivers, streams and lakes that are clean, clear and safe. This mission is achieved through many habitat restoration projects, public education, and volunteer opportunities and partnering with local businesses and organizations.

PCS is the umbrella organization for the Maumee Remedial Action Plan Committee in the Maumee Area of Concern (AOC), which is located in northwestern Ohio, US (Fig. 2.1). The Maumee AOC, which is the largest Great Lakes Area of Concern in the state of Ohio, has a complex list of water quality issues – referred to as Beneficial Use Impairments (BUIs) – which PCS works with the Ohio Environmental Protection Agency, other federal and state agencies, local municipalities, community partners, and volunteers EPA to address (Lawrence 2011). Efforts to address BUIs within the Maumee AOC have been undertaken since 1987 through various local organizations and partnerships with numerous local and regional partners and agencies, including associated programs within the State of Ohio intended to address Lake Erie (Ohio Lake Erie Commission 2013). For more information on the work of PCS in the Maumee AOC, readers are referred to: www.partnersforcleanstreams.org.

The Degradation of Aesthetics (BUI #11) has been addressed by various stream cleanup programs in the community, including a major event – *Clean Your Streams* (CYS) held each fall since 1997. CYS targets all types of trash in five major waterways including many tributaries and ditches, all within the Greater Toledo Metropolitan Area of the Maumee AOC. More than 70 volunteer groups pre-register and hundreds more volunteers "walk in" the morning of *Clean Your Streams* held each September. This includes youth groups, small businesses and large corpora-

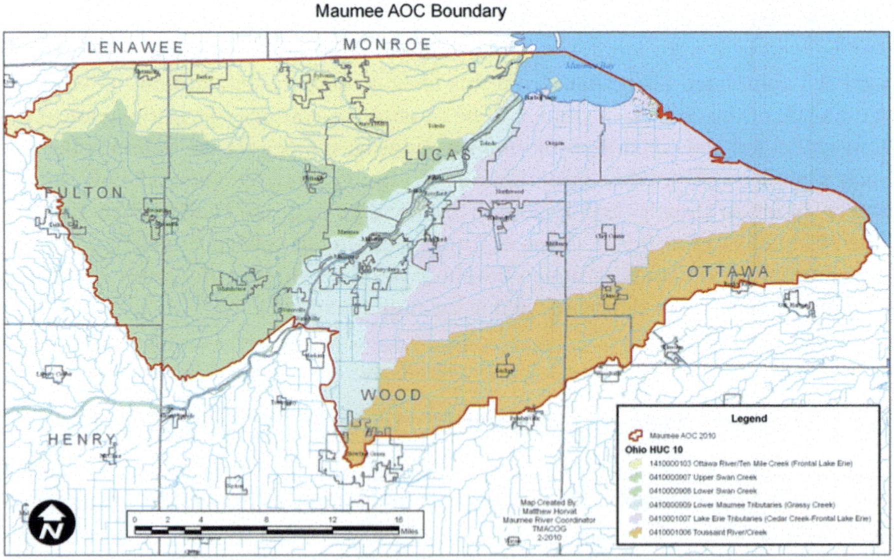

Fig. 2.1 Maumee Area of Concern, NW Ohio

tions, school groups, professional clubs, church groups, Scouts, YMCA programs, urban youth at risk groups, and various adult organizations.

For the *Clean Your Streams* program, trash is collected by different organizations, depending on location. The City of Toledo, OH and Lucas County maintenance crews donate the pick-up and deposition of collected refuse in the Hoffman Road Landfill. In and near the City of Oregon, OH, a private hauler, Waste Management, donates the hauling of the trash and disposes it at Evergreen Waste Management Landfill. Starting in 2013, all tires collected in our river cleanups were picked up by Bridgestone/Firestone contractors who worked directly with a local distributor to take the tires to a recycling facility in accordance to their "One Team, One Planet" program. More information on the annual CYS event planning and results can be found at: www.partnersforcleanstreams.org.

The *Clean Your Streams* event facilitates volunteers to remove all types of trash from local streams. In the past 17 years, thousands of volunteers have removed over 100 tons of trash from over 30 miles of ditches, streams and rivers around the Toledo, OH region. During the 2014 CYS event, 57 different streamside sites were cleaned and 16,366 lb of litter with 903 bags of trash were removed and disposed of properly. All of this litter had been disposed of improperly or illegally and through our volunteer efforts was collected and disposed of properly. Close to 1000 volunteers met for a brief introduction and safety training at 8 centralized kickoff locations and then spread out to designated sites and cleaned for about 2.5 h. After removing the often illegally dumped litter, these cleaner areas could have increased recreation use, improved water quality conditions, better aesthetics, biological productivity and reduction of waste related pollution (e.g. byproducts from breakdown of plastics). In addition to removing the trash, this program serves as an education tool to volunteers and passersby about litter. The amount of trash collected during CYS has increased from 15,315 to 18,882 lb in 2011 and 2012 respectively and 16,366 lb in 2013. In 2013, 903 bags of trash were collected, in addition to 96.5 tires and various types of "oversized" items too large for bags. These weights are estimated by our volunteers in the field and reviewed by the individual site coordinators and the event coordinator. The weights and number of bags and types of items collected are recorded on data sheets representing a team of volunteers.

Since 2013, the CYS program has significantly expanded in the number of volunteers and sites cleaned: from 726 volunteers and 40 sites to 1,175 volunteers and 60 sites being cleaned in 2011 and 2012 respectively. In 2013, there were 941 volunteers and almost 60 sites cleaned. There has been an increase of 22 % participation in the past 2 years, which is typical for the historically expanding numbers of the event. In terms of documentation, the individual data cards, which include information on the types and numbers of items collected by small working groups, are collected and annual summaries prepared for each of the five watershed to complete reporting on number of volunteers, sites worked, stream miles covered, total bags filled, and estimated weight of all bags and large items collected and removed. Each year these data are also compiled into an event report that covered the entire cleanup effort within the Maumee AOC and reported to the Lake Erie Commission office, which includes the information in the annual report for the Ohio Coastal

Weeks program (statewide stream, river and Lake Erie cleanup event in the State of Ohio 1 week each September) and the state information is submitted to the Ocean Conservancy as reporting for the International Coastal Cleanup.

2.4 Stream Clean-Up Data

Since 1987, the collection of stream cleanup data during CYS has consisted of the standard data collection form provided by the Ohio Lake Erie Commission. On this form volunteers can indicate the number of items collected by selecting from a comprehensive list of potential items covering the wide range of typical cleanup debris found during such events. They can also identify, list and count any additional items not listed as well as note any odd, peculiar or unique items. Once their cleanup is complete, the volunteers (in small teams of 3–5 people and working at assigned cleanup sites) tally their results by item and provide indication of the number of bags collected and total estimated weight of bags and any large items. These cards are then collected and the data summarized into watershed and overall reports for the entire Maumee AOC that was the focus of the cleanup effort. For example in 2014, hundreds of individually completed cards from the CYS event, were compiled to prepare a report highlighting the effort by 1,109 volunteers working at 61 sites within 4 watershed to collect over 16,000 lbs of items (same types of data also available for each of the watershed).

Historically, data on the types and numbers of items collected were not summarizes on the basis of individual sites within a watershed. During this study, each individual data card completed by volunteers within the Tenmile Creek/Ottawa River watershed during CYS over the 10 year period of the event from 1995 to 2006 was reviewed and information on the totals of individual items was assembled into a database inventory. The resulting data set included information collected for a total of 22 different collection sites (Fig. 2.2) used over the 10 year period at which over 88,000 individual items were collected and inventoried on the site data cards. A total of 20 different types of items were collected. Figure 2.3 illustrates a summary of top ten items collected, led by food wrappers and containers (18 % of total), plastic bags and wrappers (12 %), and metals cans at 12 %. Of note is that food or beverage related items, associated with fast food or grocery related products, accounted for four out of the top five items collected.

The next step in the analysis was to examine the 10 year trend in the data from the CYS event during the period 1995–2006 to see if there were any differences over time in reference to the relative amounts of different items collected. Figure 2.4 displays the trends for the top ten collected items from the Tenmile Creek/Ottawa River watershed 22 sites. Although there is considerable variability from year to year and over the 10 year record, a few general trends can be observed. The amount of food wrapper and containers (expressed as a % of the total) shows an increase from under 3 % prior to 2002, to becoming the most common item collected in 2006 at 31 % in 2006. Although not increasing as dramatically, there were also increases

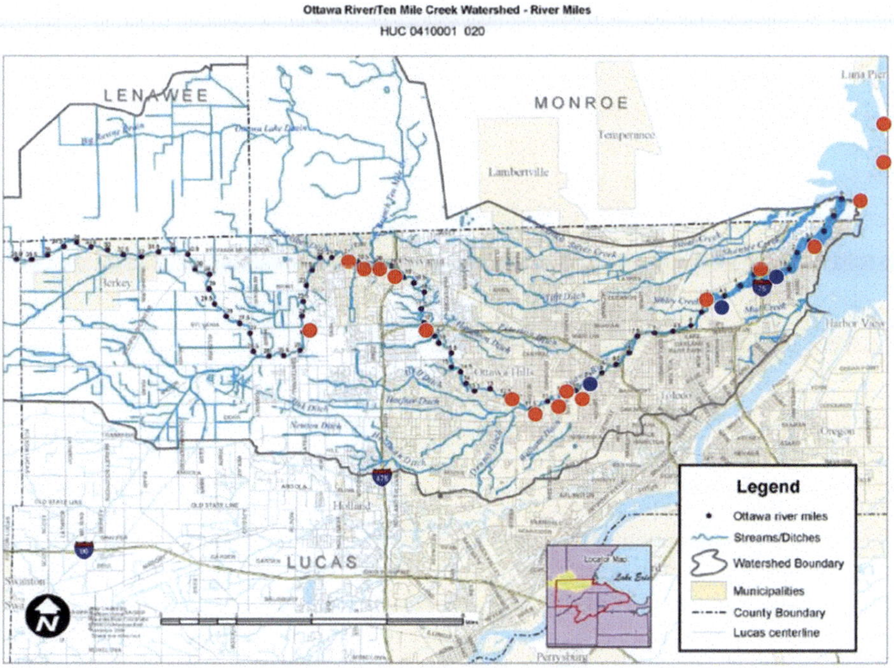

Fig. 2.2 Data collection sites from Tenmile Creek/Ottawa River

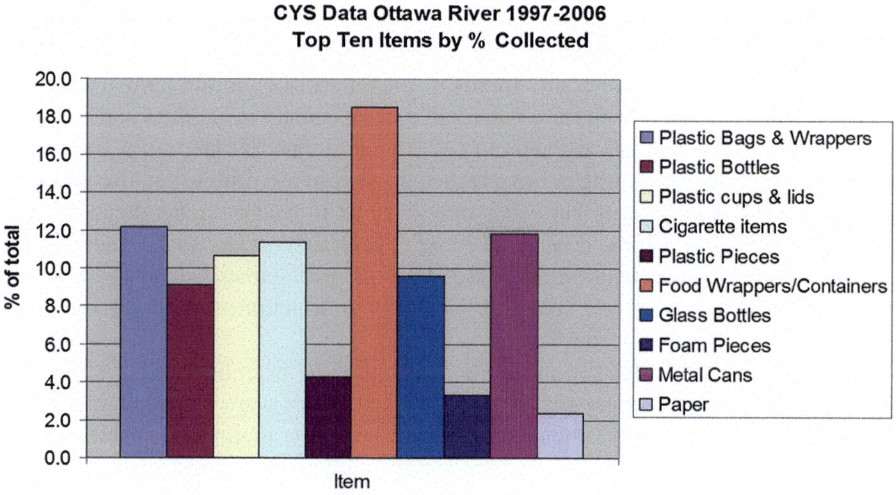

Fig. 2.3 Summary of items collected, 1997–2006

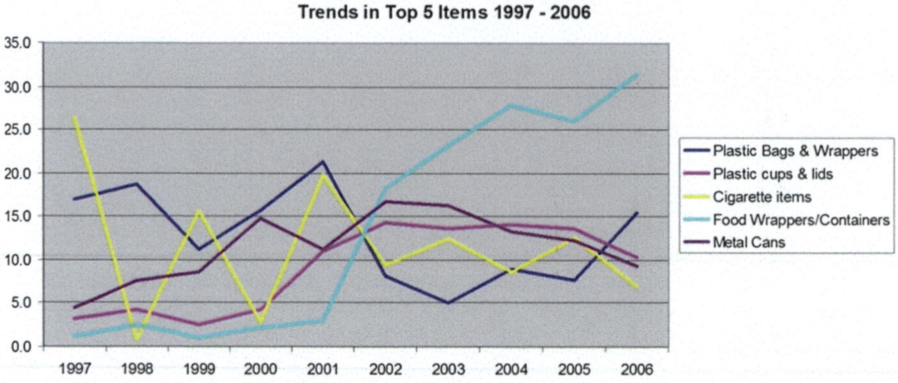

Fig. 2.4 Top 5 items collected, 1997–2006

in the number of plastic bags and cups over the decade period. However, caution should be expressed about attempting to draw to many specific inferences from this data set, since the collection of the information is based on the recording of items from a large number of differing volunteers, who although all are using the same type of data card, may not be consistent in terms of what is being collected and how it is recorded. But it is worth noting these general trends in considering future stream cleanup events within the CYS program and how the variations in items collected could inform event planning, targeted efforts at collection, and discussions regarding implementing debris sorting during CYS for separating recyclables. Such issues will be covered in more detail within the discussion section of this chapter.

Another level of basic analysis that was completed was at the site level, where for each of the 22 sites the collection data were compiled into annual calculations of total and % collected by item, and trends of the ten most common items collected. Since the sites extended from rural suburban to high density residential to mixed use with recreational (open space) lands to commercial/industrial land uses, there was interest in not only examining debris cleaned up at individual sites, but also whether differing local and proximal land uses may have an impact on items deposited (or dumped) and subsequently picked up by CYS volunteers. Decadal trends would also assist in the identification and assessment of problematic sites where items were collected year after year so as to consider various actions to stop reoccurrence of the presence of items at these sites.

In examining the collection of the most common items collected at each of the 22 sites a few interesting occurrences and general trends were observed and documented. For example, in the sites located near to roads or major street intersections more windblown and "floaters" were collected, such as paper products and plastic bags. As we move from the upstream sites and head down the river from mile 17 to mile 10 we can observe an increase in plastic bottles. In parkland and other public accessible areas, more cigarette butts are collected. Yet, food wrappers and containers (from fast food and grocery stores) remains consistently high in amounts at all

22 collection sites. Glass bottles and metal cans did not observe any spatial trends and remained in low accounts at all sites and over the entire decade collection period.

The final level of analysis involved an attempt to examine the dominant land use activities located along the stream collection sites and with the adjacent ½ mile proximity. The aim was to examine whether trends in the types and amounts of items collected at an individual stream site could be related to those human activities associated with land uses resulting in the deposit and subsequent wind or water movement of debris into the edge of stream or bank. In order to conduct this analysis a set of representative sites were examined, each of which had a distinctive set of adjacent land uses. The land uses were documented and characterized on the basis of examining recent aerial images for the sites. Sites selected included: mixed residential/commercial with parkland; floodplain area that included golf course and parkland with trails; the University of Toledo main campus; and an industrial/commercial land use site. Figure 2.5 represents the preliminary characterization of the mix of land uses located at the Ottawa Park floodplain site. The colored arrows represent conceptualization of potential major vectors for the transportation – via wind or water – from the source area (initial production or deposition) into the collection site at the stream edge. Orange represents movement off local park trails and other public access points around the park; yellow shows sources from an

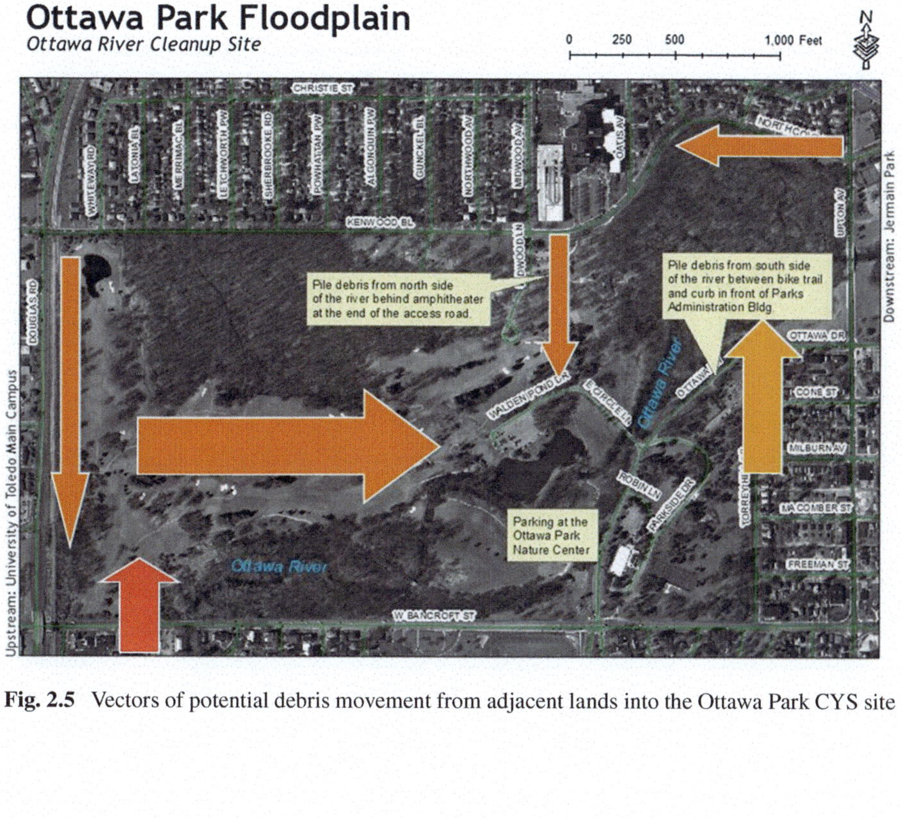

Fig. 2.5 Vectors of potential debris movement from adjacent lands into the Ottawa Park CYS site

adjacent residential neighborhood that would include homes, apartment buildings, parking lots, sidewalks, driveways and streets; while red portrays stream flow and floodplain deposition from the river into the golf course and parklands. By this means we can start to examine the collection of items from the stream cleanup activities in the context of sources and associated land use and human activities.

Also in preparing for the 2014 CYS event, the first attempt at assembling over 20 years of data on collection sites, including locations and organizing the individual site maps that had been prepared to assist volunteers in finding their assigned cleanup sites and providing them detailed information on parking, access, and where to pile their final bags with items collected for follow-up garbage hauling, was undertaken. Since the aim was to produce a consistent set of site maps (see example; Fig. 2.6) that could be readily accessible to a wide range of volunteers, rather than develop map and spatial materials by use of GIS, the decision was to prepare aerial images and resulting map products by use of Google Earth. Since 2008 over 60 individual site maps have been produced for use by event organizers and volunteers, including regular revisions and updates and adding new sites as requested by the CYS planning team members and individual kickoff coordinators responsible for site selection and allocation of volunteers across each of the five watersheds covered by the event each September.

In order to assist future CYS event planning, and to document the range and area covered by the cleanup efforts across numerous local municipalities within the five

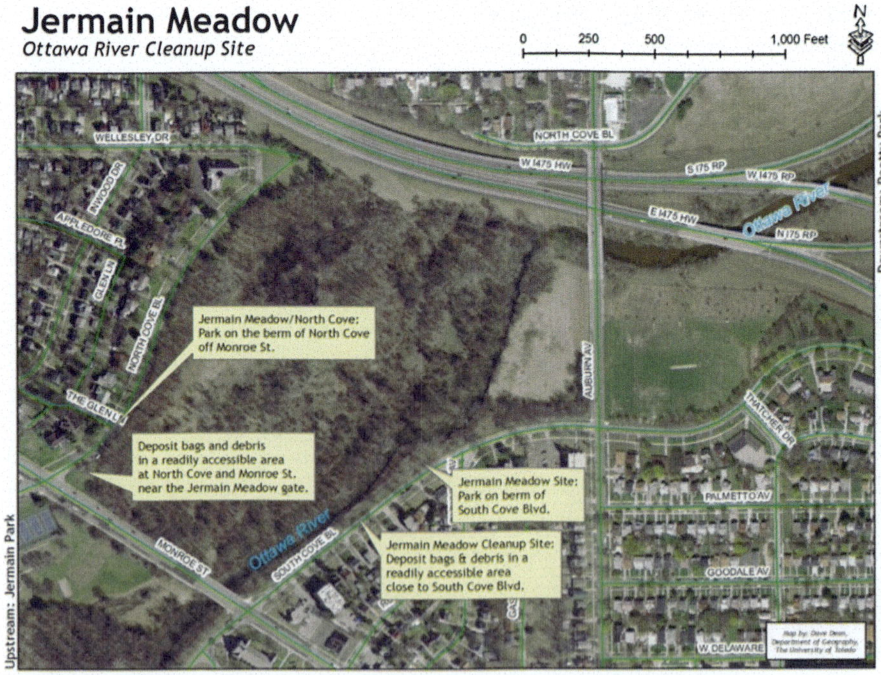

Fig. 2.6 Example of a CYS site map

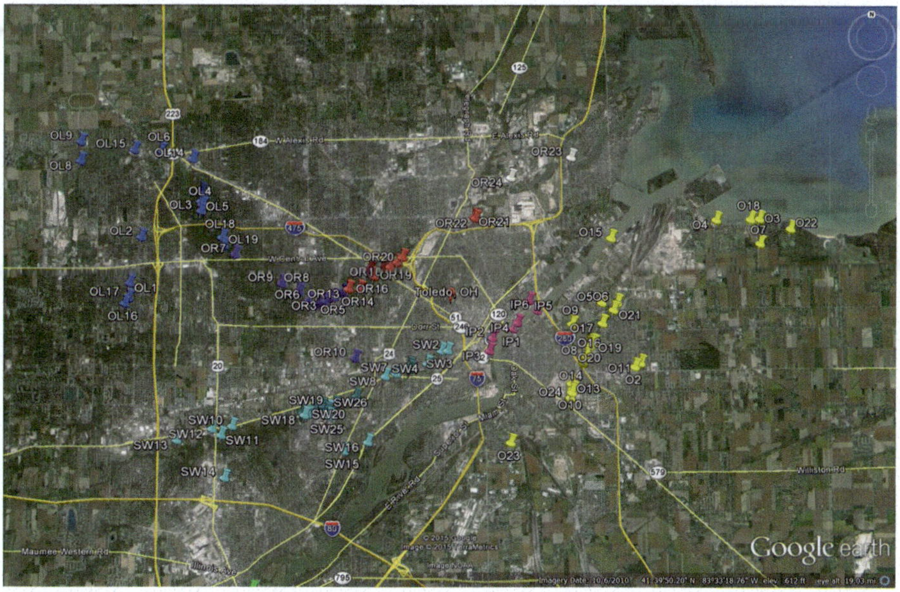

Fig. 2.7 CYS site listing map for Maumee Area of Concern

watersheds, for the first time a master CYS site map was also prepared listing by watershed every individual cleanup site used since 2000. Figure 2.7 shows the final product of this effort with over 100 sites map and color coded to each watershed, which provides a helpful understanding and illustration of the collective efforts undertaken by CYS volunteers to address stream cleanup needs within our local communities. An accompanying database was also developed that lists every site and its important characteristics and conditions, including GPS point, access, parking, stream miles to be cleaned, and where to leave debris collected for subsequent garbage pickup by municipal hauling services.

2.5 Discussion

As stream cleanup events continue to expand and increase in number both in the United States and worldwide – as an expression of the need to address restoration of urban streams and provide community based volunteer programs related to environmental concerns and issues – the ability and opportunity to move simply beyond the education value and "picking up garbage" should be expanded. Suggestions include using spatial data as means to address substantive improvements to reduce the amount of urban stream debris. As the example with the Clean Your Streams event in the Maumee Area of Concern in NW Ohio has illustrated, assembling data on locations and items collected can have additional value, especially in regards to identifying potential sources of debris from local and adjacent land uses and human activities, to noting targeted areas for continued cleanup efforts, to enacting policy and educational programs and efforts to reduce waste and collection along urban streams with communities of the Toledo OH region. Basic spatial approaches, including production of data based on location, annual and geographic trends, production of mapping of collection and site conditions, can all improve the actual stream clean up events, but also advance local and community discussions and efforts surrounding solutions and raising awareness of illicit dumping, problematic repeat offenders in terms of locations and nearby human activities that continue to generate debris, and improving public awareness around the importance of urban streams and efforts to address water quality, habitat, aesthetics and other aspects of their overall environmental conditions.

After over 20 years of the CYS program, and in the detailed examination and analysis of 10 years' worth of site level cleanup data from the Tenmile Creek/ Ottawa watershed within the Maumee AOC covered under CYS, clearly great value has been – and can continue to be – recognized by considering the temporal and spatial aspects of the distribution of debris. For example, what is being cleaned up and where, and starting to consider beyond the actual cleanup, what progress be made on addressing some of the fundamental issues related to the production and location of debris located and removed from urban streams with the local communities. Specific efforts and action items that have been undertaken as a result of CYS and reviewing results include targeted efforts to collect and safely dispose of car tires, annually reassessing cleanup site locations to target new sites that need to be addressed, discussing long term problem sites and how to better address continued debris dumping with land owners and municipal officials, and especially focusing cleanup efforts in association with areas of urban streams that have also been the location for recent natural habitat restoration projects, so that in hand with the improvement to water quality and associated environmental benefits achieved at those sites from those projects, debris removal can also add to the environmental improvements. Examples within the Maumee AOC where CYS efforts have been linked to restoration efforts include the Ottawa River at the Boy Scouts of America Camp Miakonda and University of Toledo main campus, and on Swan Creek at Highland Park.

Similar efforts and program enhancements for stream cleanup events should also be pursued at other locations, especially where well established and organized activities are already in place and supported by volunteers and organizations to facilitate such events. Especially at those stream cleanup events where data on debris collected is already recorded, for example at events conducted as part of the International Coastal Cleanup program. These programs should consider more widely how the data can be used to examine spatial context as a means to better address stream debris issues and assist with broader education and decision-making that could begin to make progress in reducing urban stream debris. Spatial analysis methods and products, as simple as basic mapping, use of GPS, developing data and event support systems based on use of geographic information system (GIS) and event applying Google Earth and other publically available, user friendly, web based or open source mapping applications and spatial tools could be of great assistance to event planning, measureable outcomes of results, and drive solutions to address and ultimately reduce urban stream debris issues within communities.

2.6 Conclusions

With the increased interests and efforts in regards to the restoration or rehabilitation of the degraded natural systems and conditions of urban streams, the use of spatial concepts and tools has great appeal to advancing our understanding of these features and considering community based decision making and planning responses to address their improvements. Rarely are urban stream ecosystems uniform in terms of their environment conditions, human impacts or potential solutions in addressing water quality, sediment, habitat, landscape functions and features, flooding and drainage, and overall beautification. It would be great benefit for those involved at the community scale in working to understand and address these issues to consider or make wider use of spatial approach tools for informational, technological, planning means to determine science and social responses and solutions.

Although stream cleanup efforts are only one approach to resolve a specific urban stream issue- the persistent presence of debris in stream ecosystems due to human activities and land uses – the simple collection and removal of such items is often the only step taken. The richness and potential application and understanding of the data of locations and items collected could greatly enhance overall efforts to improve public awareness and education efforts concerning urban streams, and spatial understanding of the debris can lead to improving cleanup events and informing as to considering solutions to address the sources of such items in order to take a more broader view of addressing debris in the context of overall stream enhancements and restoration efforts.

Urban stream cleanup programs provide great community and environmental benefits and provide a venue to advance education and public awareness on the importance and value of these often stressed ecosystems. The increased number of such events, as evidenced by the information on cleanup activities by the Ocean Conservancy with annual reporting of the International Coastal Cleanup activities worldwide, is a very encouraging and positive sign as to the rise in appreciation of urban streams within our communities and the willingness of volunteers to assist in the organization and undertaking of this effort. In many cases, the collection of relevant and important data on the debris collected can result in better understanding of the types of items ending up in urban streams and target areas and sources to of these items. In this regard it would be encouraging to see that spatial aspects be more widely considered and the use of associated mapping and analysis tools – as presented in this chapter with the CYS program in NW Ohio, US – be more widely implemented for urban stream cleanup efforts and that by doing so enhance the overall aims to improve these ecosystems in our communities and advance better understanding and appreciation of urban sustainability movements and how critical addressing the contamination and degradation of streams is within the various approaches, efforts and opportunities to restore where possible natural environmental conditions in urban settings.

Acknowledgments The Clean Your Streams annual event has been organized and conducted by numerous local groups and individuals who were responsible for conducting the event and collection of the data. These include Ohio EPA Division of Surface Water, Toledo Metropolitan Area Council of Governments (TMACOG), Partners for Clean Streams Inc., Ohio Lake Erie Commission, City of Toledo, Lucas County, Washington Township, City of Oregon, Olander Park System, University of Toledo, Village of Ottawa Hills, Toledo Metroparks, and the Boy Scouts of America Erie Shores District. Special thanks is extended to Cherie Blair (OEPA) and Kris Patterson and staff at PCS for their efforts with the annual CYS event and their assistance with the data used in this study. Appreciation is also extended to the many local community partners for providing financial support for CYS and to the thousands of volunteers who participated in the clean-up events.

References

Allison RA, Chiew FHS, McMahon TA (1998) A decision support system for determining effective trapping strategies for gross pollutants. Cooperative Research Centre for Catchment Hydrology, Clayton

American Rivers (n.d.) National River cleanup organizer's handbook. Washington, DC

Burres E (2009) Conducting rapid trash assessments. SWRCB-Clean Water Team, Sacramento

California Coastal Commission (n.d.) Municipal best management practices for controlling trash and debris. Sacramento

California Water Boards (2007) A rapid trash assessment method applied to waters of the San Francisco Bay region : trash measurement in streams. San Francisco

Friends of the Los Angeles River (2011) Final results of the first urban river challenge : Friends of the Los Angeles River vs. Friends of the Chicago River. Los Angeles

Golik A, Gertner Y (1992) Litter on the Israeli coastline. Mar Environ Res 33(1):1–15

Larson MG, Booth DB, Morley SA (2001) Effectiveness of large woody debris in stream rehabilitation projects in urban basins. Ecol Eng 18(2):211–226

Lassettre NS, Kondolf GM (2012) Large woody debris in urban stream channels: redefining the problem. River Res Appl 28:1477–1487

Lawrence PL (2011) Achieving teamwork: linking watershed planning and coastal zone management in the great lakes. Coast Manag 39:57–71

Marcus AW, Marston RA, Colvard CR, Gray RD (2002) Mapping the spatial and temporal distributions of woody debris in streams of the Greater Yellowstone Ecosystem, USA. Geomorphology 44:323–335

Maryland Department of the Environment (2010) Total maximum daily loads of trash for the Anacostia River Watershed, Montgomery and Prince George Counties, Maryland and the District of Columbia, Maryland Department of Environment, Baltimore, MD, USA

Missouri River Relief (n.d.) Big river cleanups. Columbia

Nemeth DJ, Keirsey DJ (1999) Elaboration on the nature of woody debris: an ethical snag in the aesthetic justification for organized river cleanup. Yearb Assoc Pacific Coast Geogr 61(1):86–107

New South Wales Department of Primary Industries (2007) Removal of threat abatement plan – removal of large woody debris from NSW rivers and streams. New South Wales

New York State Department of Environmental Conservation (n.d.) Removal of woody debris and trash from rivers and streams. Albany

Ocean Conservancy (2014) Turning the tide on trash. 2014 report. Washington, DC
Ohio Lake Erie Commission (2013) Lake Erie protection & restoration plan. Sandusky
Riley AL (1998) Restoring streams in cities. Island Press, Washington, DC
Stickel BH, Jahn A, Kier B (2013) Waste in our waters: the annual cost to California communities of reducing litter. Sacramento
Susquehanna River Basin Commission (1996) Managing debris in the Susquehanna River Watershed. Harrisburg
U.S. EPA (2011) Trash free waters : the urban-coastal connection. Washington, DC
Wang T (2005) Friend's of the L.A. River's trash sort. In: First state of the River report 2005. Friend's of the L.A. River, Los Angeles
Water Action Volunteers (2001) Stream and river cleanup. University Wisconsin-Extension, Madison
West Michigan Environmental Action Council (n.d.) How to do an urban streambank cleanup : the plaster creek cleanup binder. Grand Rapids

Chapter 3
Ecosystem Services Assessment from the Mountain to the Sea: In Search of a Method for Land- and Seascape Planning

Michelle E. Portman and Yarden Elhanan

Abstract Despite the growing use of ecosystem service (ES) assessment to value natural landscapes exposed to development pressures on the edge of urban landscapes, practical applications have been few. In a pilot research project, we integrate changes in different types of ES along a gradient of land- and seascape units (terrestrial, coastal and marine). This gradient includes developed urban and peri-urban areas, and undeveloped areas including submerged lands in the coastal (territorial) waters of Israel.

The emphasis in our research is on development of a practical methodology for sustainable (urban and rural) development. Goals of the pilot project presented here include: (1) evaluation of various ecosystem services based on categorical indicators for each of the four ES types: provisional, regulating, supporting and cultural; (2) development of a method that can be used on a wider scale across varying landscape units; and (3) refinement of scenario building relevant to local planning institutions and frameworks. The project's final product consists of GIS-generated maps. These maps are the basis for weighing trade-offs in ES across relevant land- and seascape units in the Mt. Carmel region of Israel. Despite some limitations, this approach to ES assessment is helpful, and if improved upon in some of the ways described herein, it can provide the first steps for infusing ES values within planning frameworks in Israel and beyond.

Keywords Ecosystems services • Israel • Environmental planning • Land use • Submerged lands • Remote sensing

M.E. Portman (✉) • Y. Elhanan
Faculty of Architecture and Town Planning, Technion-Israel Institute of Technology, Haifa, Israel
e-mail: michellep@ar.technion.ac.il; yelhanan@campus.technion.ac.il

J.D. Gatrell et al. (eds.), *Urban Sustainability: Policy and Praxis*, Geotechnologies and the Environment 14, DOI 10.1007/978-3-319-26218-5_3

3.1 Introduction

Scientists, policy-makers, and activists have promoted the ES approach to convey the extent of threats to natural ecosystems. Their goal has often been the crafting of effective and socially acceptable policy to address ecological threats. In some cases, the goal is improved decision making within a planning context (Nelson et al. 2009; Portman 2013). Environmental geography has much to offer in both cases. It can integrate between the fields of spatial ecology, geography and urban planning to support the practical application of ES as a "language" for environmental protection, conservation and sustainable development by planning professionals and stakeholders.

Practical methods for identifying, analyzing and quantifying ecosystem services (ES) on land and sea are evolving worldwide. Yet there are quite a few challenges to their implementation (e.g., Portman 2013). Incorporating cross-landscape ES value changes is identified as a challenge to the use of the ES approach in planning decision-making (de Groot et al. 2010; Kohsaka 2010). In Israel, this is especially important because planning authorities influence almost all the spatial aspects of environmental protection. Also, Israel is a relatively small, densely populated country with a heterogenous landscape. Development pressures in peri-urban areas are particularly great.

This chapter describes a case study conducted as a pilot for the Israel Ministry of Environmental Protection. It is among the few research projects examining ecosystem services along a geographic transect looking at different land and seascape units as a continuum at a local level, including areas of urban, peri-urban, marine and coastal landscapes slated for development.

Following an introduction of the concept of ecosystem services we describe challenges related to incorporating the ES paradigm for land-use planning decision making in general. We then describe the pilot study characteristics, methods and results of our research. We briefly present the project's final product, namely GIS-generated maps that indicate expected changes in ES services for the study area and a methodology that can be improved upon as new information becomes available.

3.2 Ecosystem Services: Definition and Purpose

The complex ways in which humans depend on their natural environment are increasingly assessed as ES (Adams 2014). Ecosystem services are the benefits provided to humans by natural systems which allow life and improve its quality (Reid et al. 2005). For many scientists and professionals, ES is a tool that can be used to explain to policy makers and to the general public about the importance of nature conservation and environmental protection (Carpenter et al. 2009; Collins et al. 2011; Kareiva et al. 2011; de Groot et al. 2010; Adams 2014). Some environmentalists have even described the ES approach as "the last great hope" for making biodiversity conservation a priority for planning and resource management (Daily et al. 2009).

We define ES as "the direct and indirect contributions of ecosystems to human well-being" (TEEB 2010). The ES paradigm connects, conceptually and empirically, between ecological health and human well-being. Further, it provides a way to explain and emphasize the importance of nature to policy makers and the public, using terms that make sense to scientists and laymen. Philosophically (and from a policy perspective) such advantages have utilitarian roots about values of products and processes going back to the ideas of Jeremy Bentham and John Stuart Mill (Rawls 2005).

ES are usually divided into four types (Reid et al. 2005; TEEB 2008), although recently there has been some controversy regarding the category of supporting services as explained below. The most straightforward category and perhaps the easiest one to assess is that of provisional services. Simply, these are the *products* obtained from ecosystems and as such are "services" which are directly consumed — for example, food, water, fuel, minerals, etc. "Regulating" services are those arising from ecosystems that help control environment processes (e.g., floods, erosion, climate, carbon storage, pests, pollination, waste decomposition, nutrient regulation and more (Brown et al. 2006). Cultural services are those that contribute to well-being, feelings resulting from heritage values, and from enjoyment, satisfaction, spirituality and aesthetics.

More complex is the remaining category: supporting services. These services "support" the delivery of other services, albeit often indirectly. They include such benefits as soil formation, photosynthesis, and nutrient cycling (Brown et al. 2006). Some recent studies do not include this category based on the claim that supporting services are ecological processes and therefore not ecosystem services per se (Haines-Young and Potschin 2010). Some studies (e.g., TEEB 2008) use habitat as an additional ecosystem service that supports biodiversity and forego other supports.

Measuring ES values is challenging. Problems in assessment arise firstly because ecosystem services must be actions or functions, in other words clearly valued "services". In most cases, these values are ostensibly anthropocentric. For example, consider pollination by bees. Pollination is a functional process that supports the growth of cultivated produce; while bees should be valued and are an essential part of the process, they are not the "service" itself. Rather pollination is the service. But how is it measured? By the number of bees?

A similar question can be asked about grazing areas. The service can most readily be measured according to spatial attributes such as area, continuity, vegetation type (i.e., whether or not it provides fodder), characteristics such as vegetation height, etc. But these measures are not always agreed upon. Further, valuation may depend on *which* humans are the beneficiaries (Adams 2014). Elements of the environment providing the services (such as livestock) usually have owners and this could lead to biases in assessment and evaluation. Another ambiguity relates to whether services must have some connection to the underlying ecosystem. It is problematic to assess a natural resource, especially a finite one (e.g., natural gas, sand or space), as a service because such "services" are not directly based on ecosystem functions and processes, biodiversity and/or ecosystem health. We have no shortage of these in urban and peri-urban areas where space is at a premium.

Despite these and other challenges, many countries are undertaking large-scale, national or regional-level ES assessments using various methods, categories and indicators. As examples, the UK's recently completed national ecosystem assessment (NEA 2011) classifies services similarly to the UN's Millennium Ecosystem Assessment (MEA). Finland's national ES assessment initiated in 2013 (SYKE-Finnish Environment Institute n.d.), stresses the economic importance of ES as opposed to how services contribute to nature (biodiversity) conservation. The ongoing Israel national-level assessment, considers habitat services as a category without other supporting services. "Habitat" includes breeding grounds, transfer stations for migratory species, or the protection of areas where high genetic diversity is identified (HaMaarag 2013). The creation of this separate ES category values wildlife and biodiversity more than other services.

3.3 Applying ES Assessment for Land Use Planning

Common threats to ecosystems include most land use changes and development, especially activities linked to urbanization and rapid population growth. Those most interested in ES assessment may ultimately be seeking to gauge the extent of threats and to create environmental protection and conservation policies (Stibbe 2009). Yet, ES assessment should be designed with planners and other land use professionals in mind because they are directly responsible for decisions about development, especially in urban landscapes and on the peri-urban edges, where significant ES are at stake.

In practice, the disciplinary undertaking of ES study is often biased towards ecology and economics (see Adams 2014). This poses challenges for incorporating ES within an urban planning context. Planners' decisions are not made solely on the basis of scientific data but also as a result of complex social and political considerations often related to policy (Taylor 1998; Portman 2013). Therefore, the research described herein focuses on incorporating the ES approach in a cross-disciplinary manner within an existing land-use planning context.

While many studies have encouraged integration of the ES approach in planning (e.g., Daily et al. 2009; Granek et al. 2010; Kareiva et al. 2011; Nelson et al. 2009), few have provided applicable methods for incorporating assessment results. Characteristics of local, regional or national planning frameworks vary widely from place to place which makes it hard to develop universal approaches. Some planning frameworks are characterized by decentralization; some are highly participatory and others are not. Thus pilot studies and local examples are important. These can readily be conducted using geographic information systems (GIS).

Growth in the use of GIS in the 1980s greatly improved early attempts that considered identifying the benefits of ecosystems (e.g., King 1966; Ryther 1969). GIS consolidates and combines layers of information easily which aids in the tracking of ecological processes, land use changes over time and the changes of potential planning outcomes' influence on ES. However, the use of GIS must be carefully conducted so that urban planners and decision-makers can follow the (often formulaic) modification of data. Problems frequently arise due to lack of scientific consensus regarding the models for generating GIS data (de Groot et al. 2010).

Additionally, ecosystems do not take on the boundaries of socially constructed institutions. Here we use the terms "institutions" in the broadest sense; they include reference to spatial boundaries for policy-making and planning decisions that are often incongruous to the particular extent of ecosystem types, i.e., forest, wetlands, beach, estuaries, etc. Urban master plans and regional plans used for scenario building (e.g., Nelson et al. 2009) will involve numerous ecosystem types. Broad cross-landscape unit coverage, such as that found in plans, requires comparable ES evaluations. Even when applications of ES assessments consider urban and regional plans and planning tools (e.g., Eigenbrod et al. 2010) scale can be problematic. For on-the-ground planning and decision-making, national scale assessments are often not detailed enough.

This study addresses the above-mentioned challenges. We analyze a geographic area that crosses varied land- and seascape units along a transect. As the selected transect is crossed from urban to peri-urban to rural to undeveloped, we apply different valuation methods. These are both qualitative and quantitative, as appropriate, depending on the variables we use as ES indicators.

3.4 The Mt. Carmel to the Sea Case Study

Despite the on-going preparation of National Assessment of ES for Israel (HaMaarag 2013), the country's planning institutions have yet to incorporate ES in decision-making. National Outline Plans (NOPs) such as those for parks and nature reserves (NOP 8), immigration absorption (NOP 31), and coasts (NOP 13) did not adopt the ES framework although language relating to ES is used in some (e.g., NOP 13) (Kaplan 2011). One reason is timing: the on-going Israel National ES Assessment was initiated only in 2012 while most of the national thematic outline plans are several decades old (e.g., NOS 13 was first approved in 1981, although amended since). Recognizing the potential for integration of ES assessment in work done by planning committees, the Israel Ministry of Environmental Protection has taken a lead in ES assessment and funded this work as a limited-area research pilot (Figs. 3.1 and 3.2).

The Mt. Carmel to the Sea pilot transect includes different settlement densities and types (e.g., rural and urban) of varying demographic makeup (Jewish and/or Druze) and land uses (i.e., agriculture, urban, peri-urban, beach, fishing areas). Different statutory authorities have jurisdiction within the transect, for example regional councils, the Jewish National Fund, and the Israel Nature and Parks Authority. The terrestrial area includes a range of ecosystems: forests, Mediterranean scrub, woodlands, agricultural fields, residential and developed urban areas. The coastal area includes gravel and upland dunes. The marine area stretches into the oligotrophic Levant basin to 800 m depth made up of the continental shelf's sandy and clay substrates, submerged shallow and sporadic kurkar ridges and underwater canyons.

Fig. 3.1 The national context showing the pilot transect running east to west, generally from the top of Mt. Carmel to the outer extent of Israel's territorial waters

3.5 Methods

Our work consisted of the main stages: choosing indicators from a list of ES for each of the major land- and seascape units; mapping the spatial distribution of different ecosystem services (based on the literature and available data); developing scenarios based on existing plans; estimating changes in the indicator services; estimating tradeoffs based on the indicators and developing recommendations for the

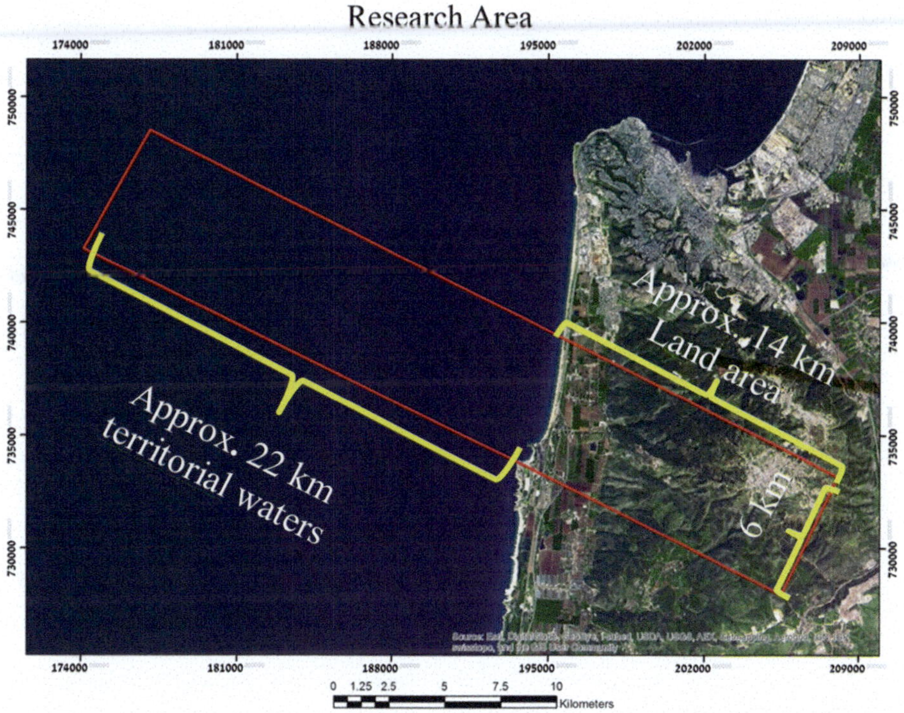

Fig. 3.2 The study area outlined in by the rectangle (with Landsat 8 as background)

future use of the methodology and its incorporation within the planning framework. We conducted some of the work simultaneously (See Fig. 3.3).

For guides, we used various reports and projects, such as the Natural Capital Project (http://www.naturalcapitalproject.org/). These provided the research team with lists from which to choose indicators (Table 3.1); the choice of indicators became a major focus of the work. Conservation planning often requires the use of indicators, due to the large numbers of plant and animal species and the lack of complete knowledge about them (Wilson et al. 2009). Also, Egoh et al. (2012) and Crossman et al. (2013) emphasize the widespread use of representative indicators, or proxies, in ES assessment according to region, land use or socio-ecological system.

Over the different landscape types (terrestrial, coastal, marine) the categories of ES allowed for some consistency. In each land- or seascape unit, we tried to choose at least one indicator for each service type: provisioning, regulating, cultural, and habitat. We based our final selection of indicators on secondary sources of GIS data and therefore were limited to three categories of ES in the marine (seascape) unit (see Table 3.2). Despite reliance in some cases on the broad ES literature, we selected indicators relevant to Israel (as determined by professional and academic

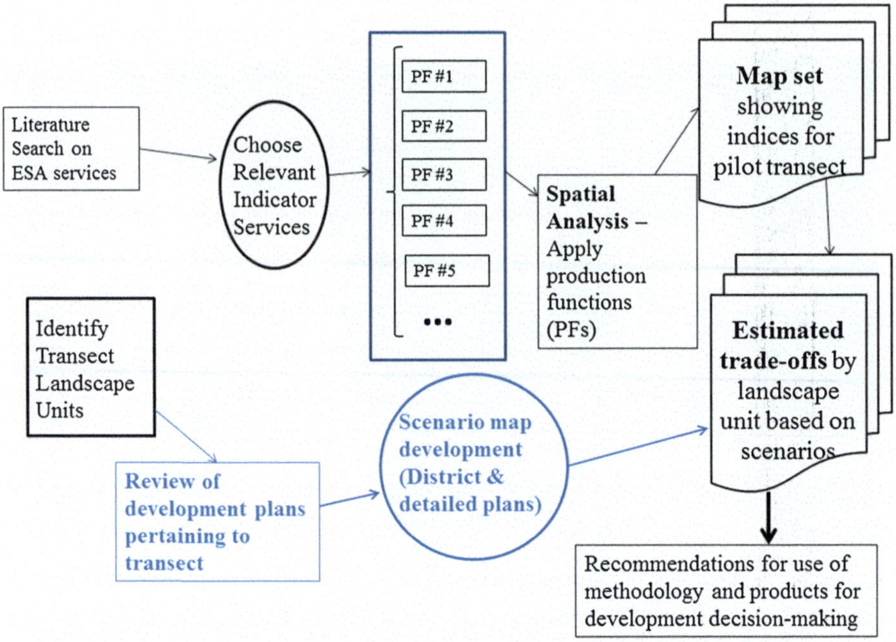

Fig. 3.3 The workflow. Note: Light blue indicates planning and policy-related action items. The concept of production functions (PFs) are explained below

Table 3.1 The ecosystem service types used according to seminal sources

MA (2005)	TEEB (2008)	Current project
Provisioning	Provisioning	Provisioning
Regulating	Regulating	Regulating
Cultural	Cultural and amenity	Cultural
Supporting	Supporting	Habitat (supporting)

literature, expert opinion and site characteristics) and those for which GIS data was readily available (e.g., Orenstein et al. 2012; HaMaarag 2013).

Combinations of fixed environmental parameters (e.g., slope and elevation) and variable environmental parameters (e.g., vegetation types and biomass) allowed calculations of indicator values. Fixed parameters change slowly (over a period of many years or longer) and include variables such as elevation, slope, ground type and more. Variable land-cover and characteristic parameters change in short periods of time and include: vegetation type, rainfall, and chlorophyll concentration in the sea.

One common variable is the Normalized Difference Vegetation Index (NDVI). The NDVI indicates whether the target area contains live green vegetation or not and to what extent. It is calculated from individual measurements of the spectral reflectance measurements acquired in the visible (red) and near-infrared regions

Table 3.2 List of terrestrial, coastal and marine ecosystem services and indices

	Service type	Value	Indices	Source(s)
Terrestrial	Regulating	Water infiltration	(NDVI × 0.5) + (Ground type × 0.3) + (slope × 0.2)	Beets et al. (2011)
	Regulating	Soil retention	(NDVI × 0.5) + (Ground type × 0.2) + (slope × 0.3)	Wylie et al. (2003)
	Regulating	Carbon storage	Changes in NDVI based on Landsat 8 images	
	Provisioning	Grazing areas	Cattle grazing ($Winter_{ndvi}$- $Summer_{ndvi}$>0.001)	
			Goats grazing ($Summer_{ndvi>}$0.2)	
	Cultural	Views (aesthetic/ landscape)	DEM layer; viewable area	Reyers et al. (2009)
			Classification area for two groups users: stationary and mobile viewers	
			Based on: height, slope, river, night camps, trail, road, picnic area, sites and land use type	
Coastal	Regulating	Water infiltration	(Soil type × 0.3) + (Vegetation density × 0.2) + (Land Use × 0.5)	Levanony (2011)
	Regulating	Erosion prevention	(Vegetation density × 0.7) + (Land use × 0.3) + (Areas within 1 km of abrasion tables)	
	Habitat	Biodiversity	Landscape value + ecological units (highest value to kurkar ridges)	Duarte (2000)
	Provisioning	Recreation	(Vegetation type × 0.5) + (0.5 × visit sites)	Barbier et al. (2011)
	Cultural	Sea views	DEM layer; viewable area	
			Based on: height, slope, river, night camps, trail, road, picnic area, sites and land use type	
Marine	Regulating	Carbon storage	(0.3 × Chloropyll) + (0.7 × depth)	Orth et al. (2006), Brown et al. (2006), and Barbier et al. (2011)
	Habitat	Marine habitats (biodiversity)	High – kurkar ridges and near shore abrasion tables and rocky outcrops	Krumbein and Van der Pers (1974)
			Medium – continental slope and canyons	
			Low – deep sea and continental slope sands	
	Provision	Water quality	According to biannual chlorophyll rates	Yahel and Angert (2012)

apparent in a satellite image. Functionally, it is the simple ratio of infrared to red characteristics of the image.

On a scale of low (barren, perhaps urban or otherwise developed areas) to high (completely forested areas), NDVI provides a measure of carbon storage taking place. NDVI values indicate about 90 % of the absorption of capacity of carbon (Beets et al. 2011; Wylie et al. 2003). A comparable variable in the ocean waters is the level of chlorophyll. In the sea, depth together with measures of chlorophyll give an indication of absorption of CO_2, in seawater, termed "blue carbon".[1]

For our study, marine GIS data was much more limited than that available for the urban, peri-urban, rural or undeveloped terrestrial and coastal areas. In the marine environment, other than levels of carbon sequestration, we used levels of chlorophyll to indicate good water quality (for purposes of desalination) and benthic information to determine habit quality.

In the coastal landscape unit, determined to be approximately up to 1 km east of the shoreline (bounded by major transportation infrastructure), we mapped vegetation types for the assessment of various ES, such as habitat. Vegetation and the existence of near-shore abrasion tables indicted land retention services (i.e., stabilization of sand dunes and coastal protection from erosion respectively). Here we also used NDVI as an indicator of carbon sequestration and sea views (cultural services) as identified through the use of a digital elevation model (DEM) GIS layer.

3.6 Building Indexes for ES Assessment

The data collected served as the basis for the production of maps indicating the current ES along the transect. For this stage we used existing land modeling techniques (such as in Reyers et al. 2009). For greater detail of variables such as slope and aesthetic views we interpolated the original 25-meter (m) DEM into a 1-m DEM for both the terrestrial landscape unit and the coastal unit. In marine areas, we used bathymetry data, which remained at the coarser 25-m resolution due to data limitations (Fig. 3.4).

We developed ES value indices (see Table 3.2) through the use of production functions (PFs) as described in Nelson et al. (2009). PFs give a value which indicates the relative benefits derived from each ES in the generic form $y=f(x)$, where y is the level of service/benefit and x is the ecosystem component or process (See the example shown in Fig. 3.5). In most cases, experts were consulted to arrive at the approximate (relative) function.

Map layers record the ES delivered for each indicator (see Fig. 3.6). Pixel values are summed between layers for an additive score. Because the set of indicators is slightly different in marine, coastal and terrestrial environments, we prepared one

[1] As opposed to black carbon (particulate matter) and brown carbon (organic aerosols), emitted respectively by the incomplete combustion of fossil fuels and biomass.

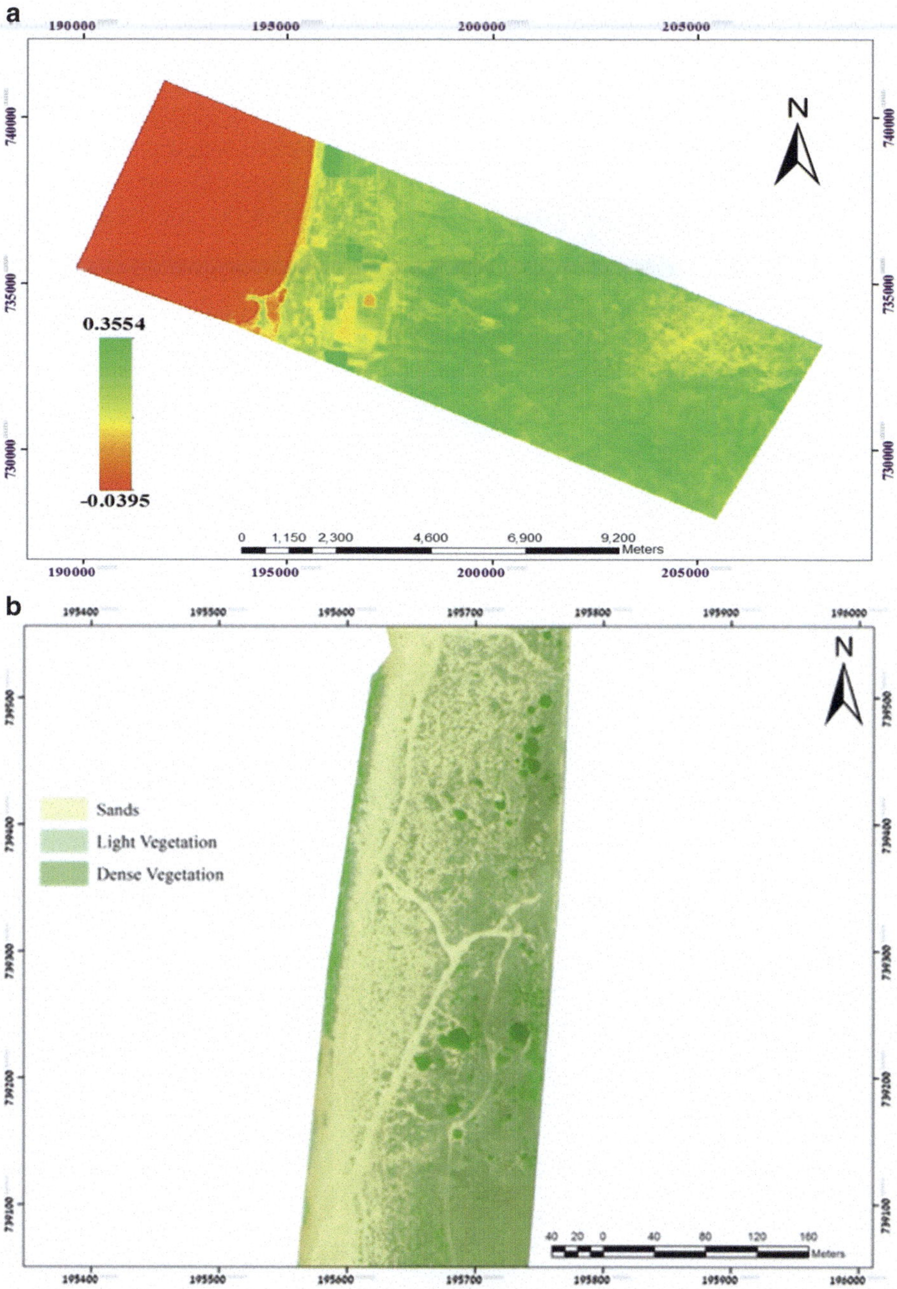

Fig. 3.4 The mapping of abiotic and biotic ecosystem components within the transect: (**a**) NDVI; (**b**) mapped coastal vegetation based on remote sensing images (LiDAR); (**c**) bathymetry which together with chlorophyll levels, indicated carbon storage services

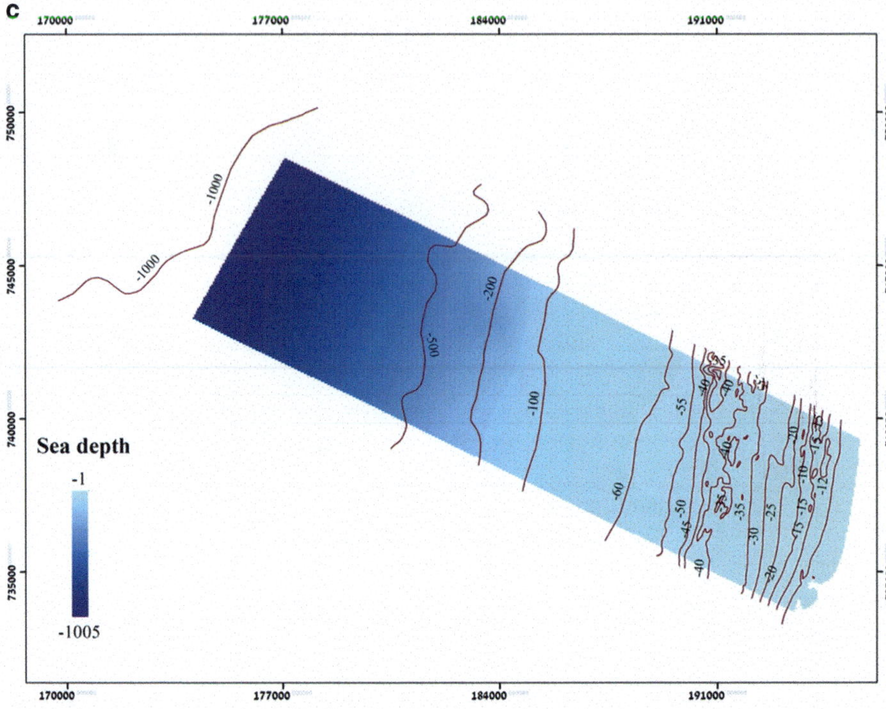

Fig. 3.4 (continued)

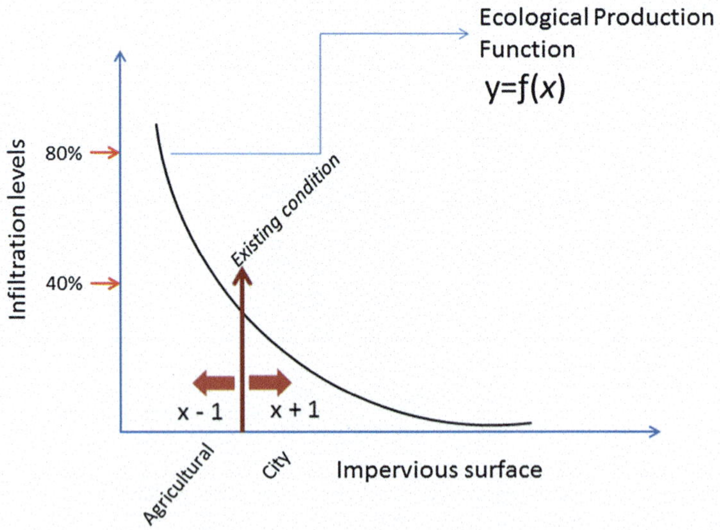

Fig. 3.5 An example PF for the service of water infiltration PFs are generalizations or models of reality (See Nelson et al. 2009)

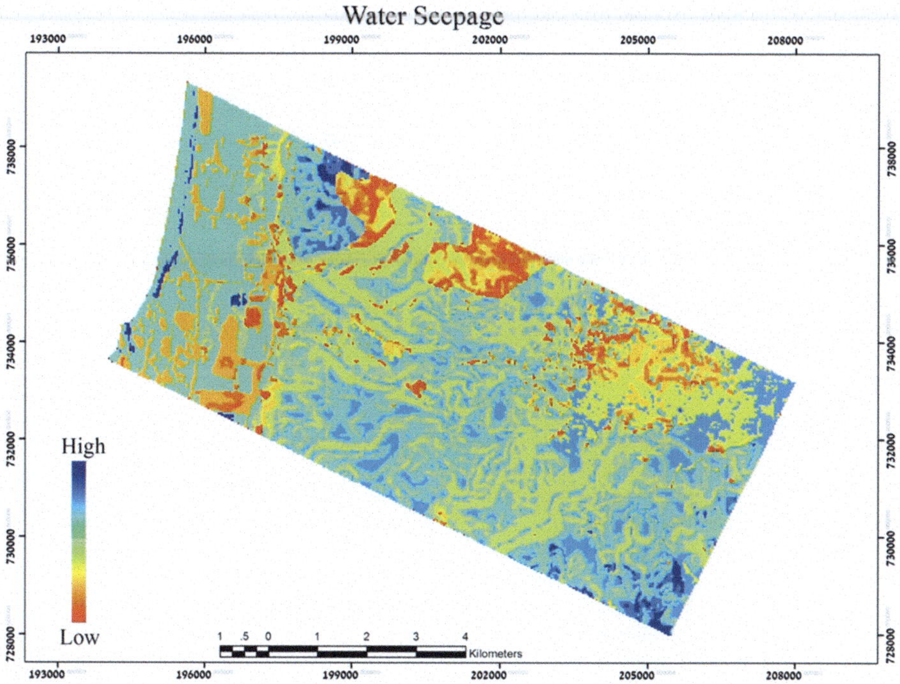

Fig. 3.6 Water infiltration visualized as a gradient, from high infiltration (*blue*) to low infiltration (*red*). The water infiltration PF shown in Fig. 3.5 theoretically describes what is seen here

composite layer for each landscape unit. These additive scores can be normalized and compared across landscape units.

Results of the entire process are the change (trade-off) in services expected from the implementation of approved plans. For the terrestrial landscape unit (Table 3.3) we devised scenarios from two sets of plans: (1) the Haifa District Outline plan; and (2) a compilation of three detailed town plans (Nir Tzion, Dalyiat El Carmel and Atlit Neighborhood Expansion). For the coastal unit we used two plans: (1) the Haifa District Outline Plan; and (2) the National Outline Plan for the Mediterranean Coast of Israel. For the marine area we used: (a) a plan for an offshore liquefied natural gas absorption station (NOS 37); and (b) a compilation of plans prepared as background for the ongoing Israel Marine Plan. As of the writing of this chapter, Israel's first marine spatial plan[2] is under development; we used that effort to cull information on expected uses of the marine area (such as areas slated for mariculture development) (Fig. 3.7).

[2] Covering approximately 27,000 km^2 of sea area. More information can be found at: http://msp-israel.net.technion.ac.il/

Table 3.3 Plans used for scenario development (only terrestrial portion)

Plan level	Land uses	Plan goals	Planning horizon
Detailed. Atlit neighborhood	Housing, public buildings, public open space, parks, commercial, infrastructure, roads, walkways	Residential sub-division; Public building/use areas; land takings	Unspecified
Detailed. Daliyat-Al-Carmel neighborhood	Housing, mixed use tourism, mixed use commercial, public buildings, public open space, infrastructure, walkways, roads	Expansion of urban area	Undefined
Masterplan. Nir Zion	Housing, public buildings, parks, roads, open spaces, hospitality	Town expansion	Undefined
Masterplan. Daliyat-Al-Carmel	Housing, public buildings, parks, agricultural, industrial, roads, open spaces, commercial, tourism, infrastructure	Setting development policies for Daliyat-al-Carmel city; emphasis on Druze identity and history	Undefined long-term
District. Masterplan Haifa District	Towns, cities, villages forests, beaches, infrastructure, agriculture	Improving: (a) commercial, social and physical development for the Haifa district, (b) urban design and construction, (c) environmental quality	2020

Fig. 3.7 The Nir Etzion (town) Masterplan. The plan shows the extent of a new neighborhood slated for construction in the Carmel Mountains

3.7 Results

We used spider graphs to show the tradeoffs expected from implementation of the development plans used for scenario building. Expected effects are shown visually against the existing ES levels for comparison and decision-making. The graphs show relative changes and make scenarios comparable. For example, as shown in Fig. 3.8, if the Haifa District Plan is fully implemented we expect significant reductions in soil control (retention) and slight reductions in grazing and infiltration services. Other services in the landscape unit will remain about the same.

Ideally, indicators can be added beyond those presented in Fig. 3.8 (water infiltration, grazing, carbon sequestration (C02), soil control) and those presented in Table 3.2. Our study used only secondary data, but many more indicators can be added from each of the service types as information improves and funding is secured for primary data collection. This holds true both for the information needed for a more accurate ES baseline assessment as well as for improved scenario building.

The study led to identification of important limitations both in terms of project and in regards to the methodology. Firstly, indicators were chosen based on availability of data and according to ES category; they are not comprehensive nor do they represent a consensus, other than that of the ES categories. Secondly, in regards to the scenarios, a binary, developed/non-developed land use change was assumed.

It is possible that some areas will deliver unanticipated ES once they are developed. For example, in the urban environment, city or neighborhood parks may still allow for a reasonable level of water infiltration. Further, urban outline plans may include parks and open spaces that will provide greater cultural values than will previously "undeveloped" land, thus contributing to urban sustainability though not necessarily as captured in ES assessment. Such analysis requires a finer scale of

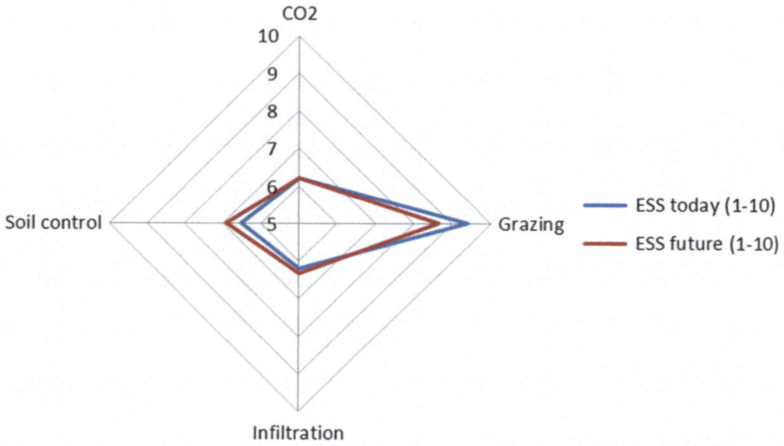

Fig. 3.8 This spider graph shows the expected effects of the Haifa District Plan on ES indicators of the terrestrial (landscape) unit. Values of one thru ten have been normalized. Such a graph was developed for each landscape unit: terrestrial, marine and coastal

information than that provided by the plans used. To conduct a full assessment of cultural values, all elements of the urban park would have to be considered. For example, an urban park may call for the construction of a paved amphitheater (without contributions to conservation) or, if designed differently, it could provide significant bird habitat, both contributing to urban sustainability, albeit in different ways.

Other limitations concern the use of various parameters such as slope, soil type and vegetation cover. As additive scoring is conducted it becomes complex and hard to follow. Having some type of automated system, or GIS application solely dedicated to conducting these analyses, would be helpful. Standards could be introduced for consistency and tracking. More research is needed in this area.

The assessment of ES loss based on approved plans served as the last step in the pilot, evidence that the change in ES can inform planning praxis and thus can contribute to sustainable development within the planning context. Ideally, the ES assessment should be done before plans have been approved and not afterward, as we did. We posit that the same or similar GIS layers of information can be used to assess change expected from future proposals or can be analyzed to identify the loss or gain in ES from different detailed plan alternatives.

3.8 Conclusions

Although the ES approach is just one tool, among many, to achieve nature conservation (Adams 2014), it is increasingly used to articulate the value of nature for humans. Integrating research on ES (of which there is no shortage) into the decision-making processes for sustainability planning remains a significant challenge to the ubiquitous use of ES assessment. This concern has been expressed widely by reports such as the Intergovernmental Platform for Biodiversity and Ecosystem Services (UNEP 2011), and by experts from the planning and conservation fields (e.g., Kareiva et al. 2011; Portman 2013; Adams 2014).

The use of advanced GIS mapping techniques and ES indicators are an important part of our analysis as are the use of production functions and scenario building based on urban and regional plans. The compilation of indices (function results) provides a ***relative*** estimation of ES to be lost through proposed development (scenarios). Scenarios can be formed from various sources: national master plans, approved district master plans and local-level detailed plans. These plan types show development intentions and can alert policy makers to potential (i.e., planned) losses in the ES that will impact sustainability. Furthermore, once the indicators have been chosen for each of the landscape units and comparisons can be made between them through normalization, trade-offs are apparent.

The contribution of this research pilot comes largely from the development of its methodology. Through it we: (1) illustrated the use of indicators for application of the ES assessment across land- and seascape units in urban, peri-urban and undeveloped areas; (2) developed maps that indicate ES values in three landscape units; and

(3) identified shortcoming in the method that can be improved upon for future use. Despite some limitations, we believe that this type of ES assessment methodology can make a significant contribution to sustainability planning and can provide the first steps for infusing ES values within the planning framework of Israel.

Acknowledgments The authors wish to thank other members of the research team for this project: Semion Polinov, Dr. Daniel Orenstein and Dr. Tally Katz-Gerro. This research is partially funded by the Israel Ministry of Environmental Protection, grant no.121-5-1.

References

Adams WM (2014) The value of valuing nature. Science 346(6209):549–551

Barbier EB, Hacker SD, Kennedy C, Koch EW, Stier AC, Silliman BR (2011) The value of estuarine and coastal ecosystem services. Ecol Monogr 81:169–193

Beets PN, Reutebuch S, Kimberly MO, Oliver GR, Pearce SH, McGaughey RJ (2011) Leaf area index, biomass carbon and growth rate of radiata pine genetic type and relationships with LiDAR. Forests 2(3):637–659

Brown C, Corcoran E, Herkenrath P, Thonell J (eds) (2006) Marine and coastal ecosystems and human well-being. United Nations Environment Programme, Nairobi

Carpenter SR, Mooney HA, Agard J, Capistrano D, DeFries RS, Diaz S, Dietz T, Duraiappah AK, Oteng-Yeboah A, Pereira HM, Perrings C, Reid WV, Sarukhan J, Scholes RJ, Whyte A (2009) Science for managing ecosystem services: beyond the Millennium Ecosystem Assessment. Proc Natl Acad Sci 106(5):1305–1312

Collins SL, Carpenter SR, Swinton SM, Orenstein DE, Childers DL, Gragson TL, Grimm NB, Grove JM, Harlan SL, Kaye JP, Knapp AK, Kofinas GP, Magnuson JJ, McDowell WH, Melack JM, Ogden LA, Robertson GP, Smith MD, Whitmer AC (2011) An integrated conceptual framework for long-term social ecological research. Front Ecol Environ 9(6):351–357

Crossman ND, Burkhard B, Nedkov S, Willemen L, Petz K, Palomo I, Drakou EG, Martín-Lopez B, McPhearson T, Boyanova K, Alkemade R, Egoh B, Dunbar MB, Maes J (2013) A blueprint for mapping and modeling ecosystem services. Ecosyst Serv 4:4–14

Daily GC, Polasky S, Goldstein J, Kareiva PM, Mooney HA, Pejchar L, Ricketts TH, Salzman J, Shallenberger R (2009) Ecosystem services in decision making: time to deliver. Front Ecol Environ 7:21–28

De Groot RS, Alkemade R, Braat L, Hein L, Willemen L (2010) Challenges in integrating the concept of ecosystem services and values in landscape planning, management and decision making. Ecol Complex 7(3):260–272

Duarte CM (2000) Marine biodiversity and ecosystem services: an elusive link. J Exp Mar Biol Ecol 250(2000):117–131

Egoh B, Drakou EG, Dunbar MB, Maes J, Willemen L (2012) Indicators for mapping ecosystem services: a review. Publications Office of the European Union, Luxembourg

Eigenbrod F, Anderson BJ, Armsworth PR, Heinemeyer A, Gillings S, Roy DB, Thomas CD, Gaston KJ (2010) Representation of ecosystem services by tiered conservation strategies. Conserv Lett 3:184–191

Granek EF, Polasky S, Kappel CV, Reed DJ, Stoms DM, Koch EW, Kennedy CJ, Cramer LA, Hacker SD, Barbier EB, Aswani S, Ruckelshaus M, Perillo GME, Silliman BR, Muthiga N, Bael D, Wolanski E (2010) Ecosystem services as a common language for coastal ecosystem-based management. Conserv Biol 24:207–216

Haines-Young R, Potschin M (2010) Proposal for a common international classification of ecosystem goods and services for integrated environmental and economic accounting (V1) (CICES). Paper presented at the fifth meeting of the UN Committee of experts on environmental-economic accounting, New York, 23–25 June 2010

HaMaarag Israel's National Ecosystem Assessment Program (2013) Report on the state of nature in the Mediterranean Sea [in Hebrew], Jerusalem

Kaplan M (2011) National outline plan for forests and afforestation: NOP 22 policy document. Keren Kayemeth L'Israel, Jerusalem

Kareiva P, Tallis H, Ricketts TH, Daily GC, Polasky S (2011) Natural capital: theory and practice of mapping ecosystem services. Oxford University Press, Oxford

King R (1966) Valuation of wildlife resources. Reg Stud 3(1):41–47

Kohsaka R (2010) Developing biodiversity indicators for cities: applying the DPSIR model to Nagoya and integrating social and ecological aspects. Ecol Res 25(5):925–936

Krumbein WE, van der Pers JNC (1974) Diving investigations on biodeterioration by sea-urchins in the rocky sublittoral of Helgoland. Helgoländer Meeresun 26(1):1–17

Levanony T (2011) Kurkar and Hamra habitats: beauty spots of the coastal plain [in Hebrew]. Society for the Protection of Nature in Israel, Tel Aviv. http://www.teva.org.il/_Uploads/dbsAttachedFiles/KurkarBooklet.pdf

Nelson E, Mendoza G, Regetz J, Polasky S, Tallis H, Cameron DR, Chan KMA, Daily GC, Goldstein J, Kareiva PM, Lonsdorf E, Naidoo R, Ricketts TH, Shaw MR (2009) Modeling multiple ecosystem services, biodiversity conservation, commodity production, and tradeoffs at landscape scales. Front Ecol Environ 7(1):4–11

Orenstein DE, Groner E, Argaman E, Boeken B, Preisler Y, Shachak M, Ungar E, Zaady E (2012) An ecosystem services inventory: lessons from the northern Negev long-term social ecological research (LTSER) platform. Geogr Res Forum 32:96–118

Orth RJ, Carruthers TJB, Dennison WC, Duarte CM, Fourqurean JW, Heck KL Jr, Hughes AR, Kendrick GA, Kenworthy WJ, Olyarnik S, Short FT, Waycott M, Williams SL (2006) A global crisis for seagrass ecosystems. Bioscience 56(12):987–996

Portman ME (2013) Ecosystem services in practice: challenges to real world implementation of ecosystem services across multiple landscapes: a critical review. Appl Geogr 45(0):185–192

Rawls J (2005) A theory of justice. Belknap Press of Harvard University Press, Cambridge

Reid WV, Mooney HA, Cropper A, Capistrano D, Carpenter SR, Chopra K, Dasgupta P, Hassan R, Leemans R, May RM, Pingali P, Samper C, Scholes R, Watson RT, Zakri AH, Zhao S (2005) Ecosystems and human well-being: a synthesis. In: Sarukhan, Whyte (eds) Millenium ecosystem assessment. World Resources Institute, Washington, DC

Reyers B, O'Farrell PJ, Cowling RM, Egoh BN, Le Maitre DC, Vlok JHJ (2009) Ecosystem services, land-cover change, and stakeholders: finding a sustainable foothold for a semiarid biodiversity hotspot. Ecol Soc 14(1):38

Ryther JH (1969) Photosynthesis and fish production in the sea. Science 166(3901):72–76

Stibbe A (2009) The language of sustainability: shouting but not being heard. Keynote address at language of sustainability: shouting but not being heard conference, Institution for Environmental Sciences, London, 27 April 2009

SYKE-Finnish Environment Institute (n.d.) National Assessment of the Economics of Ecosystem Services in Finland (TEEB Finland) – Synthesis and Roadmap. http://www.syke.fi/en-US/Research__Development/Research_and_development_projects/Projects/National_Assessment_of_the_Economics_of_Ecosystem_Services_in_Finland_TEEB_Finland/National_Assessment_of_the_Economics_of_(16944). Accessed 21 Oct 2014

Taylor N (1998) Urban planning theory since 1945. Sage, London

TEEB (2008) The economics of ecosystem & biodiversity: an interim report. Banson, Cambridge

TEEB (2010) The economics of ecosystems and biodiversity: ecological and economic foundation. Earthscan, Cambridge

UK NEA (UK National Ecosystem Assessment) (2011) The UK national ecosystem assessment: synthesis of the key findings. UNEP-WCMC, Cambridge

UNEP (2011) UNEP chief first addresses first session of IPBES plenary meeting. http://www.unep.org/Documents.multilingual/Default.asp?DocumentID=52&ArticleID=8880&l=en&t=long

Wilson KA, Cabeza M, Klein CJ (2009) Chapter 2: fundamental concepts of conservation prioritization. In: Moilenan A, Wilson KA, Possingham H (eds) Spatial conservation prioritization: quantitative methods and computational tools. Oxford University Press, New York
Wylie BK, Johnson DA, Laca E, Saliendra NZ, Gilmanov TG, Reed BC, Tieszen LL, Worstell BB (2003) Calibration of remotely sensed, coarse resolution NDVI to CO2 fluxes in a sagebrush-steppe ecosystem. Remote Sens Environ 85:243–255
Yahel R, Angert N (2012) Conservation policy for the Mediterranean Sea [in Hebrew]. Israel Nature Parks Authority, Jerusalem. http://old.parks.org.il/sigalit/yam/mediniut-shmurot-yam.pdf

Chapter 4
Best Practices for Urban Hyperspectral Remote Sensing Data Acquisition and Processing

Vijay Lulla and Ryan R. Jensen

Abstract There is an increase in use of hyperspectral remote sensing to map, monitor, and model the urban environment. This is due to the development of hardware and software systems that enable collecting hyperspectral images from aircraft based platforms. While hyperspectral data have many benefits relative to traditional multispectral data, such as fine spectral and spatial resolutions, hyperspectral data acquisition and processing is fraught with challenges that must be considered. This chapter discusses some of the common issues associated with acquiring and using hyperspectral data in urban areas. Specific issues include flight pattern planning, geometric accuracy (GPS/INS problems and square pixels), impact of aircraft flight dynamics, radiometric accuracy and correction, field data collection, data transformation techniques, data storage, and many others. The chapter also describes ways to address these issues. We think that the topics discussed in this chapter are timely because while hyperspectral remote sensing with remotely piloted aerial systems (e.g., fixed wing drones, quadcopters, octocopters) has not *yet* become popular we are already observing that these systems are being used to collect very high spatial resolution multispectral aerial imagery for many applications, and we think that hyperspectral remote sensing may become even more popular in urban areas with remotely piloted aerial systems. In this chapter, we share our experience of collecting, processing, and using hyperspectral data from aircraft platforms to study urban environments and promote urban sustainability.

Keywords Hyperspectral remote sensing • GPS/INS • AISA+ • Post processing

V. Lulla
Department of Geography, Indiana University-Purdue University, Indianapolis, Indianapolis, IN, USA

R.R. Jensen (✉)
Department of Geography, Brigham Young University, Provo, UT, USA
e-mail: ryan.jensen@byu.edu

J.D. Gatrell et al. (eds.), *Urban Sustainability: Policy and Praxis*, Geotechnologies and the Environment 14, DOI 10.1007/978-3-319-26218-5_4

4.1 Introduction

Accommodating ever-increasing populations of our cities, urban planners are continuously faced with challenges of managing and sustaining the living space within the city while providing all the necessary services to its inhabitants. This requires inventory, management, and judicious use of resources, both natural and anthropogenic, available in urban areas. City parks and urban trees/forests are examples of natural resources that are very valuable to city residents. Apart from the commonly understood benefits of urban forests such as environmental, economic, and social, they also provide habitat for birds, animals, and fish (Jensen et al. 2004).

The lessons and best practices developed from long use of remotely sensed data obtained from Earth observation satellites are not readily applicable to remote sensing of urban areas. The primary reason is spatial scale. Observing land use and land cover over a large geographic area usually means observing relatively homogeneous areas. However, urban areas are very spatially complex and there is a need for greater spatial and spectral resolutions (Small 2003, 2005). With the increasing ease and accessibility for collecting fine resolution (spatial and spectral) airborne imagery (with an airplane or a remotely piloted vehicle) imagery and imagery products are now more prevalent than ever. This chapter describes some of the factors that need to be considered to collect airborne imagery with a special focus on hyperspectral remote sensing for urban forestry. We focus on urban forestry because that is what we have most experience using and measuring. Our methods and practices have been developed by using AISA+ hyperspectral sensor ("AISA Bandage User's Manual" 2003) for studying urban forests in Midwestern (Ganesh Babu 2009; Lulla 2009, 2010), and western United States (Jensen et al. 2012a, b). Others have written on similar topics (Aspinall et al. 2002). Based on our previous work and experience, we think that hyperspectral remote sensing data can be used to promote urban sustainability provided correct practices are followed.

The factors mentioned below are applicable to all airborne remote sensing systems but are particularly important for hyperspectral remote sensing of urban areas. The chapter discusses each of the factors and summarizes how they impact the data collection, processing, and interpretation of the collected data. The factors are:

1. Flight pattern: time of day, direction of flight (effect of sun angle in to-and-from flights)
2. Geometric accuracy problems:
   (a) Instrumentation problems: GPS/INS issues
   (b) Rectangular pixel problem
   (c) Effect of flight dynamics (roll, pitch, yaw) on the collected data.
3. Geometric, radiometric, and orthorectification of images.
4. Coordinating field data collection: Primarily for spectral calibration of sensor and designing a framework for ground truthing the collected data.

5. How to use data transformation techniques developed for satellite image processing to be used for aerial imagery.
6. Considerations for data processing, cataloging, storage, and retrieval.

4.2 Best Practices for Hyperspectral Data Collection

The following are general guidelines to help instruct the acquisition, processing, and use of hyperspectral remote sensing in urban areas. There are exceptions to all of the guidelines that may inform different ways to collect and analyze data over specific urban areas.

4.2.1 Flight Pattern Considerations

While most commonly used satellite derived remotely sensed images do not have to deal with flight direction, this is one of the most important aspects for aircraft based image acquisition in urban areas (Jensen 2005). Consider, if you wanted to capture images of a particular urban area, how would you choose your flight lines? While this is a complete non-issue for remotely sensed data obtained from satellite imagery, it becomes important for aerial/airborne remote sensing systems. If you flew in one direction for 30 min, and then turn around for a pass in the opposite direction, the sun angle would be different in the next pass, and it needs to be corrected to compare the two passes. How can this be avoided? One solution is simple – instruct the pilot to fly the first pass and then fly back to the beginning of the adjacent pass to start the next flight line in the same direction. Howeve, this causes other problems. It takes equal amount of time to get back and start the flight in the same direction so you'll only get half the number of images. Also, what happens if atmospheric conditions, such as cloud cover, change? The weather becoming cloudy is not a problem per se (as long as the cloud cover is consistent throughout the areas), but if the pilot acquired data on the return flight we'd have two images (albeit with sun angle problems) instead of just one image! How do we decide what we should do? We suggest that all flight lines be acquired in the same direction (e.g., south to north, etc.) and at the same time of day with generally the same sky/atmospheric conditions.

Another question that comes up is: At what height should the aerial system be flown? While we would like very fine spatial resolution data for most urban analysis, there are physical limits that need to be considered. To get finer spatial resolution data the aircraft needs to be flown at a lower altitude. Flying at a lower altitude decreases the swath width of each flight line, and causes more flight lines to be flown. The general rule for spatial resolution is to collect data that are one half of the smallest dimension of the smallest objects or phenomena that will be identified. For example, if the goal of a mission is to identify all shrubs with a diameter of 50 cm or greater on a college campus, spatial resolution should be half of the smallest

Fig. 4.1 An AISA+ hyperspectral image acquired in Terre Haute, Indiana. The road on the left of the image should be straight, but it appears crooked because of aircraft roll. Figure 4.2 contains the corrected version of this image

dimension or 25 cm. Finally, most hyperspectral remote sensing systems are 'push-broom.' This means that the sensor requires the forward motion of the aircraft to collect images, and each image line is collected separately from the previous line. This is different from typical point-and-shoot digital cameras where the shutter is opened, and an array of cells (charge-coupled device) is populated with electromagnetic energy. Because of this, when acquiring data with a hyperspectral sensor, aircraft roll greatly affects the field of view of the sensor and tilts the sensor to either the port or starboard side of the aircraft (Fig. 4.1).

Unless overlap is planned between flight lines, aircraft roll will cause missing data throughout the imaged area. We suggest that at least 20 % overlap be planned for the flight lines. Greater overlap should be planned if the area is prone to thermal updrafts. Table 4.1 shows how aircraft height above ground level impacts pixel resolution and swath width with relation to scan rate (speed that the sensor moves to the next line) and aircraft speed.

Table 4.1 The relationship between aircraft height above ground level, aircraft speed, sensor scan rate, pixel size, and image swath width using an AISA+ hyperspectral scanner

Flight height		Airplane speed (km/h)				Sq Pix size	
(Feet)	(Meters)	Scan rate 30	Scan rate 31	Scan rate 32	Scan rate 33	(Meters)	Swath (m)
2000	610	93	96	99	102	0.861	441
2500	762	116	120	124	128	1.076	551
3000	914	139	144	149	153	1.29	661
3500	1067	163	168	174	179	1.506	771
4000	1219	186	192	198	204	1.721	881
4500	1372	209	216	223	230	1.937	992
5000	1524	232	240	248	256	2.152	1102
5500	1676	256	264	273	281	2.366	1211
6000	1829	279	288	297	307	2.582	1322
6500	1981	302	312	322	332	2.797	1432
7000	2134	325	336	347	358	3.013	1543
7500	2286	349	360	372	383	3.227	1652
8000	2438	372	384	396	409	3.442	1762

4.2.2 Geometric Accuracy Considerations

For high spatial resolution hyperspectral data to be useful for any sort of analysis it needs to be geometrically accurate. Images are geometrically accurate when they align with corresponding features on the ground. Most aerial hyperspectral systems use GPS/INS (Geographic Positioning System/Inertial Navigation System) to record geographic coordinates and flight characteristics while the images are being acquired. Most researchers are well aware of GPS receivers – including how they work, what data they collect, and how they can be used. However, INSs are not as widely known. An INS continuously records the position, orientation, and speed of the moving system with reference to some starting point. INS data are commonly used as navigation aids, but we need INSs in airborne hyperspectral imaging so that we can remove errors due to flight dynamics (roll, pitch, and yaw). In our experience the GPS/INS has been the most problematic area of such an undertaking (Jensen et al. 2008, 2011). We understand that by the very nature of the function that INSs perform, errors will have a cumulative effect but these errors can be so prominent so as to make any meaningful analysis impossible. System vendors usually provide software utilities that use the recorded GPS/INS system to georectify the images but they are useless if the unit did not function correctly. We have consistently had difficulty with this and we have corrected many images using some ad-hoc procedures such as straight-line correction where image geometry is forced to follow a straight line in the image, such as a road, to correct aircraft roll error (Jensen et al. 2008, 2011; Fig. 4.2).

Fig. 4.2 This image of Terre Haute, Indiana was corrected using the straight-line correction method

While we have been able to fix these issues we strongly believe that these work-arounds should be unnecessary. We're very hopeful that widespread use of navigation sensors (primarily in smartphones and consumer electronic devices) and a growing interest in remotely piloted vehicles will help alleviate this problem.

4.2.3 Radiometric Considerations

Even though using hyperspectral systems for collecting data has the potential to solve many of our problems because their fine spatial and spectral characteristics, these data need to be radiometrically corrected before they can be used for any sort of image processing analysis. The common problems that need to be rectified are geometric rectification (as described above – problems due to flight dynamics of the aerial system), and atmospheric correction (influenced by atmospheric conditions) (Mäkisara et al. 1994a, b).

For atmospheric corrections there are model-based and empirical correction methods. Some of the commonly used model-based methods for atmospheric correction are Atmospheric REMoval program (ATREM), Atmospheric CORrection

Now (ACORN), and the Fast Line-of-sight Atmospheric Analysis of Spectral Hypercubes (FLAASH) (Kruse 2004). Commercial vendors of aerial hyperspectral imaging systems usually provide in-built software that allows for atmospheric rectification. For example, the AISA+ hyperspectral imaging system comes with an atmospheric correction module as a part of the package ("AISA Bandage User's Manual" 2003). This module is called 'Caligeo' and it is installed as an add-on the image processing package called ENVI (Environment for Visualizing Images). To effectively use Caligeo, additional field data collection is required. The implications of this are discussed in the next section.

4.2.4 Field Data Collection

Regardless of what kind of remotely sensed imagery/data are used for analysis, it is imperative to have corresponding field data because field data are needed for additional radiometric calibration of the sensor, image classification development, accuracy assessment (Congalton and Green 1999) of the classification, and model generation (relationship between biophysical variables and reflectance). Radiometric calibration is achieved by recording the reflectance values of plastic tarps (generally 2×2 m) of known reflectance characteristics when the aerial system is flying overhead. In addition, this also provides a ground control point in the imagery. For this to be useful there needs to be at least a few more such measurements in different parts of the flight coverage area and in all flight lines. This necessitates multiple field personnel and all the logistics and planning that need to go into coordinating this field measurement.

More field data collection (on the same day of the flight) is needed if you plan to develop a model between biophysical variables and hyperspectral data. Developing a relationship between urban leaf area index (LAI) and hyperspectral data is an example of such modeling (Jensen et al. 2012b). This requires considerable planning and careful organization obtain all the necessary field data that allows effective use of urban hyperspectral remote sensing data. For example, ideally ground measurements are collected along each flight line while the aircraft is overhead acquiring the images. This becomes very difficult when there are many flight lines (e.g., >10) and limited people and equipment are available for ground measurements.

4.3 Best Practices for Hyperspectral Data Processing

4.3.1 Data Transformation Considerations

The sheer amount of data (e.g., number of bands) that are collected in hyperspectral remote sensing poses problems to effectively use the data. For example, Table 4.2 lists the center wavelengths (nanometers) of all 248 bands available when using the AISA+ hyperspectral sensor.

Table 4.2 Center wavelengths (nanometers) of all 248 bands in the AISA+ hyperspectral sensor. Bands in this configuration are 2.3 nm wide

Band	Wavelength (nm)	Band	Wavelength (nm)	Band	Wavelength (nm)
1	391.99	84	577.72	167	772.07
2	394.17	85	580.04	168	774.45
3	396.35	86	582.37	169	776.83
4	398.53	87	584.69	170	779.21
5	400.71	88	587.02	171	781.59
6	402.89	89	589.34	172	783.97
7	405.06	90	591.67	173	786.35
8	407.24	91	593.99	174	788.73
9	409.42	92	596.31	175	791.10
10	411.60	93	598.64	176	793.48
11	413.78	94	600.96	177	795.86
12	415.96	95	603.29	178	798.24
13	418.14	96	605.61	179	800.62
14	420.31	97	607.94	180	803.00
15	422.49	98	610.26	181	805.38
16	424.67	99	612.58	182	807.76
17	426.85	100	614.91	183	810.13
18	429.03	101	617.23	184	812.51
19	431.21	102	619.56	185	814.89
20	433.39	103	621.88	186	817.27
21	435.56	104	624.21	187	819.65
22	437.80	105	626.53	188	822.03
23	440.04	106	628.85	189	824.41
24	442.29	107	631.18	190	826.78
25	444.53	108	633.50	191	829.16
26	446.77	109	635.83	192	831.54
27	449.02	110	638.15	193	833.91
28	451.26	111	640.48	194	836.29
29	453.50	112	642.80	195	838.66
30	455.74	113	645.12	196	841.04
31	457.99	114	647.45	197	843.41
32	460.23	115	649.77	198	845.79
33	462.47	116	652.10	199	848.17
34	464.72	117	654.42	200	850.54
35	466.96	118	656.75	201	852.92
36	469.20	119	659.07	202	855.29
37	471.45	120	661.39	203	857.67
38	473.69	121	663.72	204	860.04
39	475.93	122	666.04	205	862.42
40	478.18	123	668.37	206	864.80
41	480.42	124	670.69	207	867.17

(continued)

Table 4.2 (continued)

Band	Wavelength (nm)	Band	Wavelength (nm)	Band	Wavelength (nm)
42	482.66	125	673.02	208	869.55
43	484.91	126	675.34	209	871.92
44	487.15	127	677.66	210	874.30
45	489.39	128	679.99	211	876.67
46	491.64	129	682.31	212	879.05
47	493.88	130	684.64	213	881.43
48	496.12	131	686.96	214	883.80
49	498.36	132	689.28	215	886.18
50	500.61	133	691.61	216	888.55
51	502.85	134	693.93	217	890.93
52	505.09	135	696.26	218	893.30
53	507.34	136	698.62	219	895.68
54	509.58	137	700.98	220	898.06
55	511.82	138	703.35	221	900.43
56	514.07	139	705.72	222	902.81
57	516.31	140	708.08	223	905.18
58	518.55	141	710.45	224	907.56
59	520.80	142	712.81	225	909.93
60	523.04	143	715.18	226	912.31
61	525.28	144	717.54	227	914.69
62	527.53	145	719.91	228	917.07
63	529.77	146	722.28	229	919.45
64	532.01	147	724.64	230	921.83
65	534.25	148	727.01	231	924.21
66	536.50	149	729.37	232	926.60
67	538.74	150	731.74	233	928.98
68	540.98	151	734.10	234	931.36
69	543.23	152	736.47	235	933.74
70	545.47	153	738.84	236	936.12
71	547.76	154	741.20	237	938.50
72	550.06	155	743.57	238	940.88
73	552.37	156	745.93	239	943.26
74	554.67	157	748.30	240	945.64
75	556.98	158	750.67	241	948.02
76	559.28	159	753.04	242	950.40
77	561.59	160	755.42	243	952.78
78	563.89	161	757.80	244	955.17
79	566.20	162	760.18	245	957.55
80	568.50	163	762.56	246	959.93
81	570.81	164	764.94	247	962.31
82	573.11	165	767.32	248	964.69
83	575.42	166	769.70		

While there are methods of selecting hyperspectral bands (Bajcsy and Groves 2004), determining what bands or band combinations will be useful will depend on the area of study and what is being studied. For example, there are different kinds of reference spectral libraries (Baldridge et al. 2009) of all the features occurring on the surface of the Earth that should be consulted to determine which are the most optimal bands for a particular area. Further, anyone who has worked with NDVI (Normalized Difference Vegetation Index) probably remembers that NDVI requires measurements (bands) acquired in both infrared and red portions of electromagnetic energy and is an indicator of vegetation health. Selecting these bands is quite simple when there is only one band in each portion of energy. However, band selection becomes more difficult when the infrared and red portions of the electromagnetic spectrum have tens of bands in them. Which combinations should be used? The short answer is that the selection of bands to calculate NDVI and other vegetation indices is dependent on the specific sensor data (Jensen et al. 2012b).

We have experience using AISA+ hyperspectral data in urban settings. This sensor collects data in hundreds of spectral bands (Table 4.1). Creating statistical models that use this as input data needs to account for interaction between these bands (variables in a case by variable terminology commonly used in statistics). Or it requires reducing the number of bands by some data reduction technique like Principal Component Analysis (Jolliffe 2004), or use other techniques such as artificial neural networks (Benediktsson et al. 1990; Kavzoğlu 2001; Lulla 2010).

4.3.2 Storage and Retrieval Considerations

In recent years data storage has gotten much cheaper and faster, and data storage is no longer a limiting factor with large datasets. For example, external hard drives with three terabytes of storage can now be purchased for around $200. However, cataloging and retrieving data are altogether different matters. There are still not many easy to use programs/systems to catalogue and index stored data. Of course, data stored on network servers such as Google, or Amazon's S3 (Simple Storage Service) are indexed and easy to search but those companies have spent a lot of research and development resources to create the infrastructure that makes this possible. Currently, no such option is available for local storage. Moreover, data retrieval is primarily based on querying very simple text attributes of the data (commonly called metadata). While this is simple, it requires planning and effort to keep it all coordinated and manageable.

Anyone who has tried to obtain satellite based remotely sensed data from many of the online data portals is aware of the workflow of selecting images that meet the user's criteria (for example, select a date range, sensor type, geographic coordinates, cloud cover percentage, etc.). All of this is possible because the data portal service has cataloged/indexed these data and the data now have an interface that the user can query. If an organization decides to collect its own hyperspectral data, it will have to replicate a similar system, albeit at a much smaller scale, to query and retrieve pertinent data. Also, the organization will need to develop some sort of naming conventions for different types of files (Abrams and Hook 2002). All these rules apply not only to the raw acquired data but also to any preprocessing that has been performed on these data.

Data cataloging could also help to query images/data based on some intrinsic characteristics. For example, land use and land cover (LULC) classified images are commonly used for different types of environmental and ecological modeling. Assume an analyst has classified all the hyperspectral images into such LULC images. Now if we want to find all classified images that meet a particular classification criteria (20 % of one class and 28 % of another class), how do we do it? Also, how do we incorporate classification accuracy into our decision making process? What about differences in classification accuracies of surrounding images? Is this even possible (maybe it's not possible due to flight considerations discussed earlier)? Unfortunately, we do not have reasonable answers for these questions, and future efforts in data cataloging and query may attempt.

We are excited by some of the advances in database technologies that might be beneficial to make urban hyperspectral more accessible (manageable) to more people. One such development is the rise in popularity of array databases that provide database services for raster data. With this rise in popularity of scientific software there is more awareness of multidimensional arrays and matrices amongst the programming community. This has sparked interest in embedding something similar in databases that will allow for easy access and querying. SciDB and PostGIS are a couple of examples that we're aware of and we expect to see more such offerings and integrations.

4.4 Conclusions

We have described some of the factors that impacted much of the urban hyperspectral remote sensing work that we have done in the Midwestern and western United States. The development in recent years, of hyperspectral technologies, widely, and sometimes freely (Biehl and Landgrebe 2002), available image processing software, and access to relatively inexpensive cloud computing resources (Amazon's ElasticCompute, Google App Engine, and Microsoft Azure) excites us about the continued use of hyperspectral remote sensing in urban areas for a variety of applications. We believe that researchers who implement best practices for acquiring and analyzing hyperspectral data can effectively use these data to better understand the urban environment study and promote urban sustainability.

References

Abrams M, Hook S (2002) ASTER user handbook – version 2. Jet Propulsion Laboratory/EROS Data Center, Pasadena

AISA Bandage User's Manual (2003)

Aspinall RJ, Marcus WA, Boardman JW (2002) Considerations in collecting, processing, and analysing high spatial resolution hyperspectral data for environmental investigations. J Geogr Syst 4(1):15–29

Bajcsy P, Groves P (2004) Methodology for hyperspectral band selection. Pers 70(7):793–802

Baldridge AM, Hook SJ, Grove CI, Rivera G (2009) The ASTER spectral library version 2.0. Remote Sens Environ 113(4):711–715

Benediktsson JA, Swain PH, Ersoy OK (1990) Neural network approaches versus statistical methods in classification of multisource remote sensing data. IEEE Geo 28(4):540–552

Biehl L, Landgrebe DA (2002) MultiSpec – a tool for multispectral – hyperspectral image data analysis. Comput Geosci 28:1153–1159

Congalton RG, Green K (1999) Assessing the accuracy of remotely sensed data: principles and practices. Lewis Publications, Boca Raton

Ganesh Babu B (2009) Classification of urban features using airborne hyperspectral data. Dissertation, Indiana State University, Terre Haute, Indiana

Jensen JR (2005) Introductory digital image processing: a remote sensing perspective. Prentice Hall, Upper Saddle River

Jensen R, Gatrell J, Boulton J, Harper B (2004) Using remote sensing and geographic information systems to study urban quality of life and urban forest amenities. Ecol Soc 9(5):5, [online] http://www.ecologyandsociety.org/vol9/iss5/art5

Jensen RR, Jackson MW, Lulla V (2008) Single line correction method to remove aircraft roll errors in hyperspectral imagery. J Appl Remote Sens 2:023529

Jensen RR, Hardin AJ, Hardin PJ, Jensen JR (2011) A new method to correct pushbroom hyperspectral data using linear features and ground control points. GI Sci Remote Sens 48(3):416–431

Jensen RR, Hardin PJ, Hardin AJ (2012a) Classification of urban tree species using hyperspectral imagery. Geocarto Int 27(5):443–458. doi:10.1080/10106049.2011.638989

Jensen RR, Hardin PJ, Hardin AJ (2012b) Estimating urban leaf area index (LAI) of individual trees with hyperspectral data. Photogramm Eng Remote Sens 78(5):495–504

Jolliffe IT (2004) Principal component analysis. Springer, New York

Kavzoğlu T (2001) An investigation of the design and use of feed-forward artificial neural networks in the classification of remotely sensed images

Kruse FA (2004) Comparison of ATREM, ACORN, and FLAASH atmospheric corrections using low-altitude AVIRIS data of Boulder. In: Paper presented at the 13th JPL airborne geoscience workshop, JPL Publication 05–3, Pasadena, CA

Lulla V (2009) Hyperspectral applications in urban geography, Planning and socioeconomic applications. Springer, Dordrecht

Lulla V (2010) Neural network classification of hyperspectral imagery for urban environments: a case study. Dissertation, Indiana State University, Terre Haute, IN

Mäkisara K, Kärnä J-P, Lohi A (1994a) Geometric correction of airborne imaging spectrometer data. In: Paper presented at the international geoscience and remote sensing symposium; surface and atmospheric remote sensing; technologies, data analysis, and interpretation, Pasadena, CA

Mäkisara K, Lohi A, Kärnä J-P (1994b) A system for geometric and radiometric correction of airborne imaging spectrometer data. In: Paper presented at the international geoscience and remote sensing symposium; surface and atmospheric remote sensing; technologies, data analysis and interpretation, Pasadena, CA

Small C (2003) High spatial resolution spectral mixture analysis of urban reflectance. Remote Sens Environ 88(1–2):170–186

Small C (2005) A global analysis of urban reflectance. Int J Remote Sens 26(4):661–681

Chapter 5
Thermal Neighborhoods, Socioeconomic Characteristics, and Urban Quality of Life: Examining Humanity's Principal Habitat

Ryan R. Jensen and Perry J. Hardin

Abstract Urban heat can negatively impact urban quality of life – a measure of resident well-being. However, the burden of urban heat is often unevenly distributed. This chapter examines urban temperature in Terre Haute, Indiana using detailed land cover data and socioeconomic data. First, four urban thermal neighborhoods were generated using land cover percentages and k-means clustering. Then, urban temperature was examined in the context of these four neighborhoods, and it was found that urban temperature was significantly different in the four neighborhoods. Finally, socioeconomic variables were examined. It was determined that income was significantly higher in neighborhoods with low temperatures and percentage minority population values were significantly higher in the neighborhoods with the higher temperature values. Future research could further investigate the relationship between urban heat and socioeconomic characteristics.

Keywords Urban heat island • Urban quality of life • Socioeconomic characteristics • Orthophotos • ASTER surface kinetic temperature

5.1 Introduction

As humanity's principal habitat, urban areas are important for a number of reasons. For example, urban areas provide services, job opportunities, and entertainment that surrounding rural areas usually cannot provide. Urban areas are said to exist because of cities are the most economically efficient structure for to distribute relationships among individuals (Royuela et al. 2010). Also, people tend to socialize and congregate in places where they're most comfortable, and urban areas can provide a sense of protection and safety. Finally, much of the rural areas surrounding urban centers

R.R. Jensen (✉) • P.J. Hardin
Department of Geography, Brigham Young University, Provo, UT, USA
e-mail: ryan.jensen@byu.edu

J.D. Gatrell et al. (eds.), *Urban Sustainability: Policy and Praxis*,
Geotechnologies and the Environment 14, DOI 10.1007/978-3-319-26218-5_5

are being converted to more developed land as urban areas continue to expand outward. Indeed, the vast majority of the world's cities are growing at an ever-increasing rate as a result of birth rates higher than death rates and migration from rural areas.

Urban areas also have negative features such as, congestion, increased crime rates, and little or no nearby nature (Royuela et al. 2010). Other negative features of urban areas include increased air and other kinds of pollution and diffusion of disease (de Hollander and Staatsen 2003). Given both the positive and negative urban features, it is important to study the urban environment and its associated biophysical and social characteristics – especially the quality of human life in urban areas and features that impact this quality of life.

5.1.1 Urban Quality of Life

Urban quality of life (UQL) is a relatively new urban subject that gained momentum in the 1980s by initiating new forms of environmental rules and regulations, and introducing alternative strategies and tactics to mainstream environmentalism (Floyd and Johnson 2002; Jensen et al. 2004). After President Clinton's Executive Order 12898 (United States Government Archives 1994), the issue of environmental equity and justice in natural resource allocation received increased attention (Tarrant and Cordell 1999). Since then significant research has been completed to address these issues, and UQL has become a primary focus within environmental social science (Weinberg 1998). Furthermore, UQL has become a chief focus for planners, funding agencies, and local communities (Steinberg 2000). Improving urban quality of life is an important goal of public policy that goes beyond the traditional provision of social services (Santos and Martins 2007). As an economic factor, UQL is often used to attract the attention of people looking to locate or relocate in urban areas – people often compare the bundle of amenities available at one location to those offered in another location (Blomquist et al. 1988).

Urban quality of life studies have used geographic techniques, such as geographic information systems (GIS) and remote sensing, to analyze urban conditions and illuminate core policy issues and stratagems to improve the material conditions of everyday life, and the overall quality of life of residents (e.g., Porter and Tarrant 2001; Harner et al. 2002; Mennis 2002). Pedlowski et al. (2002) examined the relative wealth, species diversity, and abundance of urban forests in Brazil. Although not definitive, their analysis suggested a relationship between class and urban forest amenities. Rinner (2007) examined UQL using geographic visualization techniques. The author found that the method provided an interactive view that supplied data on spatial patterns that could be used in making informed decisions.

Unfortunately, one of the difficulties in defining UQL is developing methods to measure and weigh the amenities provided in a particular area (Blomquist et al. 1988). To address this issue, this chapter examines UQL through the prism of urban land cover, urban heat, and socio-demographic characteristics. The chapter is built on two working hypotheses. The first hypothesis is that detailed land cover characteristics can be used to classify urban neighborhoods and that these neighborhoods will have different levels of urban heat. The second hypothesis is socio-demographic characteristics will also be different in the neighborhood classes – thus signifying that the burden of urban heat is not distributed evenly across the socio-demographic range.

This chapter examines these hypotheses in Terre Haute, Indiana using several steps. First, detailed land cover was digitized using high-resolution aerial photographs in 377 locations throughout Terre Haute. Then, these land cover characteristics were used to classify different physical neighborhood classes. Next, the temperature was measured for each of the points, and the average Celsius values were calculated for each class. Between-class means were then compared. Finally, socio-economic characteristics of each neighborhood were compared to determine whether the burden of urban heat is equitably shared.

5.1.2 Remote Sensing of the Urban Heat Island and Urban Quality of Life

Urban areas have long been known to be hotter than surrounding rural environments, a phenomenon often referred to as the urban heat island (UHI) effect. This increase in temperature is the result of many things – both physical and biological – that either contribute to or mitigate the heat. For example, urban heat typically increases with the presence of impervious surfaces such as concrete, asphalt, and roofing shingles – among many others. Conversely, urban heat is mitigated through the presence of trees, shrubs, grasses, and reflective surfaces (Jensen et al. 2010). Most urban areas contain a complex mixture of both impervious and vegetation surfaces. This complex matrix helps to determine the amount of heat in urban environments. The ability to map and monitor urban heat dynamics is a necessary step to determine the specific role that urban land cover has in urban heat.

Many UHI studies have been conducted using remote sensing data and techniques. Lo et al. (1997) found a strong relationship between the amount of vegetation present and irradiance recorded by the Advanced Thermal and Land Applications Sensor (ATLAS) in an urban area. This same concept was reinforced by Quattrochi and Ridd (1998), who found that vegetation – especially trees – have a significant mitigating effect on thermal radiation and urban temperature. Conversely, Stathopoulou et al. (2007) established that a negative UHI Heat effect (i.e., urban land is cooler than adjacent rural land) could occur during the daytime because of the differing heating properties of each surface. The negative UHI effect was also noted by Xian and Crane (2006), who examined the UHI in both Tampa, Florida, and Las Vegas, Nevada, using Landsat data. They found that urban Las Vegas exhibits a cooling effect whereas Tampa Bay exhibits a more typical urban heating effect. However, in both cities, areas with higher percentages of impervious surfaces were usually associated with higher temperatures. Both of these studies underscore the value of measuring precise land cover quantities when investigating the UHI.

Other studies have examined the specific impact of vegetation characteristics on urban temperature. A sample of this work follows:

- Hardin and Jensen (2007) found that a biophysical measure, Leaf Area Index, accounts for much of the variation in urban surface kinetic temperature.
- Weng and Yang (2004) examined urban heat and city growth. While urban expansion increases urban temperature, those increases can be mitigated by strategically planting vegetation.
- Chen et al. (2006) observed that rapidly urbanized areas are more prone to the increases in temperature than areas of slow development. The authors observe that this may be the result of initial vegetation clearing and subsequent planting of immature grass, shrubs, and trees.
- Kottmeier et al. (2007) examined the urban microclimate and the role impervious and vegetated surfaces have in contributing to surface temperature. They concluded that the cooling effect from shading (primarily from trees, high-growth vegetation, and buildings) may outweigh the heating effect of rooftops.
- Katpatal et al. (2008) found a strong relationship between a Landsat TM derived land cover dataset and urban meteorological data. They also suggested that remote sensing data can be used to accurately depict the relationship between land use/cover and urban temperature.
- Peña (2008) used Landsat Enhanced Thematic Mapper Plus (ETM+) to investigate the correlation of urban surface temperature and urban land cover to explain the creation of an urban "heat sink" in Santiago, Chile.
- Xiao et al. (2007) found that the spatial arrangement of impervious surfaces was highly correlated with land surface temperature in Beijing, China.
- Liu and Weng (2009) investigated the relationship between urban temperature and land surface patterns at multiple spatial scales in Indianapolis, Indiana. They found that 30 m was the optimal resolution for assessing the relationship between land use and land surface temperature.

While much has been done to study urban forest and the urban heat island, only a few investigators have treated it as a matter of environmental justice. For example, Heynen et al. (2006) found that the urban forest in Milwaukee was not distributed equally between all socioeconomic groups. The authors concluded that this was a violation of urban environmental equality since the urban forest contributes to urban quality of life.

5.2 Methods

5.2.1 Study Area

Terre Haute, Indiana is the government seat of Vigo County. Terre Haute is located in west-central Indiana along the Wabash River and has a population of 60,785 (United States Census 2000). Terre Haute is typical of urban areas with a heterogeneous landscape consisting of impervious surfaces (e.g., asphalt, concrete, shingles, etc.), vegetation, bare soil, and other surfaces. The city of Terre Haute is set amidst a larger landscape of agricultural fields and deciduous hardwood forests. There is very little urban development between Terre Haute and larger cities in the region (Hardin and Jensen 2007).

5.2.2 *Land Cover Mapping*

Land cover data were acquired from leaf-on digital orthophotos produced by the United States Geological Survey. The true-color orthophoto imagery data were acquired during summer 2003 with a spatial resolution of 1 m. Land cover classes were determined in 377 random locations using on-screen digitizing techniques. The following land cover classes were manually identified and digitized from the orthophotos: water, grass, rural pasture or crops (no bare soil), trees and shrubs, bare soil, mostly bare soil (some rural vegetation), mostly rural vegetation (some bare soil), light paving, dark paving, light roof, and dark roof. These 11 land cover classes were inclusive of all the land cover types present in Terre Haute. Each of the 377 locations was identified by a center point latitude and longitude. Land cover percentages in 55 m around each point (110 m quadrats) were digitized and noted for the above classes. This quadrant size was selected because 110 m was determined to be the optimum scale for mapping and monitoring urban heat using ASTER thermal data (Jensen et al. 2010).

5.2.3 *Remote Sensing Temperature Data*

Thermal infrared data acquired from the Advanced Spaceborne Thermal Emission and Reflection Radiometer (ASTER) were used to calculate surface kinetic temperature (SKT). The ASTER remote sensing device operates on the TERRA Satellite and measures reflected energy in two portions of the electromagnetic spectrum – visual near infrared (VNIR; four bands; 15 m) and shortwave infrared (SWIR; six bands; 30 m). It also measures emitted thermal infrared information (TIR; five bands; 90 m). The SKT data product (Data Product 8) is derived from these five thermal bands. The data product includes both the surface temperature and emissivity for each pixel within an ASTER scene (Hardin and Jensen 2007).

Pu et al. (2006) found SKT data derived from the ASTER sensor to be very reliable According to Gillespie et al. (1999), each of ASTER's thermal bands has a radiometric accuracy of 1 K and precision of <0.3 K. The ASTER data used for this chapter were acquired on 2 July 2001 at 1100 local time and had a spatial resolution of 90 m.

5.2.4 K-Means Clustering

As a statistical method, Cluster analysis attempts to define clusters that are relatively homogeneous within clusters and heterogeneous between clusters. In other words, data cases with similar characteristics will be grouped together, and cases with dissimilar variable values would be placed in different clusters. In this research, land cover percentages were used to determine four clusters using the k-means cluster algorithm in SPSS (Statistical Package for the Social Sciences). The k-means classifier requires two passes through the data. The first pass calculates a user-specified number of class means whereas the second pass assigns each case to one of the classes using a minimum distance to mean classification technique. For this study, four clusters were calculated using the percentages of land cover determined by the process described above. After each case was assigned to a class, general class descriptions were made to determine the land cover characteristics in each grouping.

5.2.5 Socioeconomic Data

Socioeconomic data – median income and minority population percentage – were extracted from Block Group level data acquired from the 2000 United States Census (US Census 2000). In this chapter, minority population percentage was defined as all races other than white/Caucasian. These other races included (as defined by the United State Census Bureau) black, Asian, American Eskimo, Hispanic, multi-race, and other (encompassing all other races). The sum of these races was divided by the total population in the Block Group to calculate percent minority population.

5.2.6 Statistical Comparison

Celsius values for the cases in each cluster were averaged together and compared to the average Celsius values in the other clusters using Analysis of Variance (ANOVA) to determine if there was a statistical difference in temperature between them. In a similar manner, median income and minority population percentage were averaged and compared between each cluster using ANOVA. In each of the ANOVAs, Tukey's post-hoc test was performed if the ANOVA showed a significant difference between all of the clusters to determine specific pair-wise differences.

5.3 Results and Discussion

5.3.1 *Four Classes and Temperature*

The first neighborhood class (rural; n = 55) was dominated by rural pasture or crops. This class had an average Celsius value of 30.14. The second class (paved; n = 45) contained more light paving than the other land cover types, and it was the hottest of the classes with an average of 35.16 °C. The third class (densest vegetation; n = 136) had more trees and shrubs than any other land cover. It was the coolest class with an average temperature of 28.51 °C. The last class (moderate vegetation; n = 141) contained a large percentage of grasses, trees, and shrubs. This class had an average temperature of 32.00 °C. As expected, the class with the highest amount of paving had the highest temperature. The lowest temperature class was the 'densest vegetation' class. This class contained the highest percentage of trees and shrubs.

5.3.2 *Socioeconomic Variables*

Income and housing valued were averaged and compared for each of the four classes. Results show that those with lower incomes have the highest urban heat values (paved median income = $24,157). As noted above, this class contained the highest contributions of light paving. Indeed, impervious surfaces may be indicative of lower income. The class with the lowest temperature value (densest vegetation) also had the highest median income ($46,860). The second lowest temperature class (rural) had the second highest median income ($40,402). A one-way ANOVA showed that there is a significant difference in median income between the groups (alpha <0.05). Tukey's post-hoc analysis showed that the significant differences were between 'paved' and 'moderate vegetation' and between 'paved' and 'densest vegetation.'

Minority population percentage was also different in the four classes. The highest minority percentage was in the 'paved' class (15.8 %), followed by the 'moderate vegetation' class (14.0 %), 'densest vegetation' class (8.2 %), and 'rural' class (8.1 %). The two hottest classes were 'paved' and 'moderate vegetation'. ANOVA indicated that there are differences between the groups. Post-hoc analysis showed that the minority population percentage differences were between classes 'paved' and 'moderate vegetation,' 'paved' and 'densest vegetation,' 'paved' and 'moderate vegetation,' and 'densest vegetation' and 'moderate vegetation.' As a matter of equity, this suggests that minority populations in Terre Haute are shouldering more of the burden of urban heat than are non-minorities.

5.4 Conclusions

This chapter has demonstrated the following:

- Using the methodology described above, Terre Haute neighborhoods can be grouped into four different classes predicated on the percentage of the neighborhood covered by different land cover types.
- These neighborhoods exhibit different urban heat and socio-demographic characteristics.
- In particular, median income values and minority population percentage were significantly different in the four neighborhoods – with lower income and higher minority population neighborhoods having higher average values of urban heat. As mentioned above, this may indicate that minorities and those with lower median income values unfairly bear more of the burden of urban heat than others.

As discussed at the beginning of this chapter, UQL is the agglomeration of a suite of benefits that impact urban residents – both positively and negatively. As demonstrated repeatedly, and confirmed in this work, those areas with lower incomes and higher minority populations do not have the same quality of life enjoyed by people in areas with higher incomes and lower percentages of minorities.

Based on these results, future policy makers and researchers can further explore whether the ecological and economic impacts of urban heat are evenly or unevenly distributed within and between spaces and groups. This is an especially poignant question, as it may help to determine whether a city's efforts to improve services and amenities are unfairly skewed to those areas that are 'better off' than other areas. The answers to this and other related questions are important as policy makers, citizen groups, and researchers seek to explore the true dynamics of environmental resource allocation (Environmental Protection Agency 2002).

This chapter has operationalized the general hypothesis that urban heat is significantly different in areas with varying socioeconomic conditions. As such, urban amenity variables can be used to explain and investigate the uneven distribution of urban resources. Admittedly, we do not know whether high LAI values preceded higher housing values and higher incomes, or whether the opposite is true. Furthermore, other socio-demographic variables were not analyzed. Future research could focus on analyzing different variables through time with temporal remote-sensing data and past censuses.

Finally, understanding and exploring the social and physical dynamics of urban heat are essential – the United States' urban forests occupy roughly 3 % of the country's total land area and serve as habitat for approximately 80 % of the human population (United States Census Bureau 2000; Jensen et al. 2004). Hence, it is valuable to study the distribution of urban heat within and between communities and their individual neighborhoods. With so much importance being given to habitat quality for other species in public and policy discussions, it is useful to examine the environment in which humans live as well, and determine factors that may positively or negatively influence our urban habitat (Jensen et al. 2004).

References

Blomquist GC, Berger MC, Hoehn JP (1988) New estimates of quality of life in urban areas. Am Econ Rev 78:89–107

Chen X, Zhao H, Li P, Yin Z (2006) Remote sensing image-based analysis of the relationship between urban heat island and land use/cover changes. Remote Sens Environ 104:133–146

de Hollander AEM, Staatsen BAM (2003) Health, environment and quality of life: an epidemiological perspective on urban development. Landsc Urban Plan 65:53–62

Environmental Protection Agency, Office of Environmental Justice (2002) Environmental Justice (Online). http://www3.epa.gov/environmentaljustice/resources/publications/nejac/integration-ej-federal-programs-030102.pdf

Floyd MF, Johnson CY (2002) Coming to terms with environmental justice in outdoor recreation: a conceptual discussion with research implications. Leis Sci 24:59–77

Gillespie AR, Rokugawa S, Hook SJ, Matsunaga T, Kahle AB (1999) Temperature/emissivity separation algorithm theoretical basis document. National Aeronautics and Space Administration, Goddard Space Flight Center, Greenbelt, Maryland. Version 2.4

Hardin PJ, Jensen RR (2007) The effect of urban leaf area on summertime urban surface kinetic temperatures: a Terre Haute case study. Urban For Urban Green 6:63–72

Harner J, Warner KW, Pierce J, Huber T (2002) Urban environmental justice indices. Prof Geogr 54:318–331

Heynen N, Perkins HA, Roy P (2006) The political ecology of uneven urban green space – the impact of political economy on race and ethnicity in producing environmental inequality in Milwaukee. Urban Aff Rev 42:3–25

Jensen R, Gatrell J, Boulton J, Harper B (2004) Using remote sensing and geographic information systems to study urban quality of life and urban forest amenities. Ecol Soc 9:5 [online] http://www.ecologyandsociety.org/vol9/iss5/art5/

Jensen RR, Hardin PJ, Curran R, Hardin T (2010) The relationship between urban land cover and surface kinetic temperature: a case study in Terre Haute, Indiana. J Terr Obs 2(2):45–65

Katpatal Y, Kute A, Satapathy D (2008) Surface- and air- temperature studies in relation to land use/land cover of Nagpur urban area using Landsat 5 TM data. J Urban Plan Dev 134:110–118

Kottmeier C, Biegert C, Corsmeier U (2007) Effects of urban land use on surface temperature in Berlin: case study. J Urban Plan Dev 133:128–137

Liu H, Weng Q (2009) Scaling effect on the relationship between landscape patter and land surface temperature: a case study of Indianapolis, United States. Photogramm Eng Remote Sens 75:291–304

Lo CP, Quattrochi DA, Luvall JC (1997) Application of high resolution thermal infrared remote sensing data and GIS to assess the urban heat island effect. Int J Remote Sens 18:287–304

Mennis J (2002) Using geographic information systems to create and analyze statistical surfaces of population and risk for environmental justice analysis. Soc Sci Q 84:281–297

Pedlowski MA, Da Silva VAC, Adell JJC, Heynen NC (2002) Urban forst and environmental inequality in Campos dos Goytacazes, Rio de Janeiro, Brazil. Urban Ecosyst 6:9–20

Peña MA (2008) Relationships between remotely sensed surface parameters associated with the urban heat sink formation in Santiago, Chile. Int J Remote Sens 29:4385–4404

Porter R, Tarrant MA (2001) A case study of environmental justice and federal tourism sites in southern Appalachia: a GIS application. J Travel Res 40:27–40

Pu R, Gong P, Michishita R, Sasagawa T (2006) Assessment of multiresolution and multi-sensor data for urban surface temperature retrieval. Remote Sens Environ 104:211–255

Quattrochi DA, Ridd MK (1998) Analysis of vegetation within a semi-arid environment using high spatial resolution airborne thermal infrared remote sensing data. Atmos Environ 31:19–33

Rinner C (2007) A geographic visualization approach to multi-criteria evaluation of urban quality of life. Int J Geogr Inf Sci 21:907–919

Royuela V, Morena R, Vaya E (2010) Influence of quality of life on urban growth: a case study of Barcelona, Spain. Reg Stud 44:551–567

Santos LD, Martins I (2007) Monitoring urban quality of life: the Porto experience. Soc Sci Indic 80:411–425

Stathopoulou M, Synnefa A, Cartalis C, Santamouris M, and Akbari H (2007). A heat island study of Athens using high-resolution satellite imagery and measurements of the optical and thermal properties of commonly used building and paving materials. 2nd PALENC Conference and 28th AIVC conference on building low energy cooling and advanced ventilation technologies in the 21st century, vol 2. pp 1016–1020

Steinberg MW (2000) Making sense of environmental justice. Forum Appl Res Public Policy 15:82–89

Tarrant MA, Cordell HK (1999) Environmental justice and the spatial distribution of outdoor recreation sites: an application of geographic information systems. J Leis Res 31:18–34

United States Census Bureau (2000)

United States Government Archives (1994) Federal actions to address environmental justice in minority populations and low-income populations. http://www.archives.gov/federal-register/executive-orders/pdf/12898.pdf. Last accessed 26 Jan 2015

Weinberg AS (1998) The environmental justice debate: new agendas for a third generation of research. Soc Nat Resour 11:605

Weng Q, Yang S (2004) Managing the adverse thermal effects of urban development in a densely populated Chinese city. J Environ Manag 70:145–156

Xian G, Crane M (2006) An analysis of urban thermal characteristics and associated land cover in Tampa Bay and Las Vegas using Landsat satellite data. Remote Sens Environ 104:147–156

Xiao R, Ouyang Z, Zheng H, Li W, Schienke EW, Wang X (2007) Spatial pattern of impervious surfaces and their impacts on land surface temperature in Beijing, China. J Environ Sci 19:250–256

Chapter 6
Sustainability, Greenspace and Nature Deficit in Las Vegas, Nevada

Daniel D. McLean, Ryan R. Jensen, and Elizabeth Barrie

Abstract Nature deficit occurs when people, especially children, spend less time outdoors. Nature deficit often results in a wide range of behavioral issues, and people living in urban areas are more apt to have nature deficit. The literature suggests potential relationships of nature deficit to commitments toward sustainability. Expanding the awareness of the perceived connectivity of greenspaces, nature deficit, and attitudes about sustainability are important. Geospatial technologies, such as Geographic Information Systems and remote sensing, can accurately map green spaces in urban and suburban areas. This chapter describes some of the tools and techniques used to map these areas. Mapping green spaces in urban areas can be as simple as setting a threshold of vegetation index values to determine greenspace. In this chapter we use remote sensing data and techniques to map out vegetation and greenspaces throughout the Las Vegas metropolitan area in 1985 and 2014. Specifically, the percentage of greenspace in Las Vegas has decreased by about 2.5 % in 29 years. Studies like this can help planners, geographers, and others to understand the spatial dimensions of urban greenspace, and help them to promote vegetation throughout entire urban areas – especially those areas that are deemed to not have adequate greenspace. However, further research needs to be conducted on how residents in desert areas, like Las Vegas, perceive nature (it is green?), how the desert urban landscape influences people, the contributions of desert landscape exposure to perceptions of environmental sustainability, and potential outcomes from contact with nature different in desert environments.

Keywords Nature deficit • GIS • Green spaces • Landsat • NDVI

D.D. McLean (✉)
William F. Harrah College of Hotel Administration, University of Nevada, Las Vegas, Las Vegas, NV 89154, USA
e-mail: daniel.mclean@unlv.edu

R.R. Jensen
Department of Geography, Brigham Young University, Provo, UT, USA
e-mail: ryan.jensen@byu.edu

E. Barrie
Online Education, The Office of Academic Affairs, University of Nevada, Las Vegas, Las Vegas, NV 89135, USA
e-mail: elizabeth.barrie@unlv.edu

J.D. Gatrell et al. (eds.), *Urban Sustainability: Policy and Praxis*,
Geotechnologies and the Environment 14, DOI 10.1007/978-3-319-26218-5_6

6.1 Introduction

With the publication of "Last Child in the Woods" (2005) Richard Louv issued a call to arms for American society to reconnect children with the natural world. In the wake of this call the absence of nature in the lives of American children has been broadly discussed in the research and public literature (e.g., Churchman 2003; Maller et al. 2008; Moore and Marcus 2008; Taylor et al. 2001; Turner et al. 2004). The issue took hold across the United States and politicians reacted by introducing the No Child Left Inside Act in 2007. The act consistently makes progress through house and senate subcommittees, even securing sufficient votes for passage in the House in 2008 (http://edlabor.house.gov/markups/fc20080618.shtml retrieved 7/31/2008). From 2009 through 2014 the bill was reintroduced repeatedly, but the congressional stalemate thwarted passage.

Even in the absence of legislative action, there remains a strong consensus that children of all ages are becoming increasingly disconnected from nature and sustained outdoor activity (Dustin et al. 2009; Godbey 2009). The lack of opportunities for children as well as adults to experience nature is multi-faceted, partly a consequence of the increasingly urbanized U.S. population where the percentage of people living in rural areas has dropped from 94.9 % in 1790 to nearly 15 % in 2014 (http://www.ers.usda.gov/topics/rural-economy-population/population-migration.aspx retrieved 12/29/14). This increase in urbanization removes people from easy access to natural settings (i.e., nearby nature).

6.1.1 Factors Influencing Nature Deficit

Louv (2005) suggested several factors that influence the amount of time children spend in nature including:

- The explosive growth of electronic media
- The longer work and commuting hours of parents and increased amount of scheduled activities for children that create time constraints
- The explosive rate of land development in the past two decades and the corresponding lack of nearby nature (i.e., vegetation and open space with natural features in close proximity to urban residents) in the developed areas.
- The increasingly litigious nature of American society creating risk-averse managers and citizens who prevent or limit various nature experiences (e.g., climbing trees, building tree forts, etc.)
- The prevalence of neighborhood covenants that place severe limitations on what, where, and when children can play outdoors
- The climate of fear generated by intense media attention of child abduction cases

We believe that the amount of time that people spend outdoors in nature is related to the amount of greenspaces close to their homes, and that this warrants more detailed study. Further, we contend there is a relationship between contact with nature and the development of attitudes towards environmental sustainability. Louv (2005) addressed the importance of unstructured time in nature as a palliative course of action for those in nature deficit. He asserted that unstructured time in nature is of paramount importance in combating nature deficit. Kaplan and Austin (2004) have shown that individuals consistently identify five main categories of nearby nature: manicured/landscaped areas (urban horticulture), trees, gardens, mowed areas, forest, open fields, and wetlands. The presence of these types of nearby nature suggests that there may be various types of treatments for nature deficit, creating a continuum from unstructured time in nature, structured time in nature, unstructured time outdoors (e.g., at a playground, on a ball field), and structured time outdoors (e.g., on a sports team).

The concept of nature deficit crosses many disciplines. From positive psychology to environmental psychology to resource management; and recreation to geography to education. Urban planning is reconceptualizing the location and design of public spaces but has yet to fully incorporate all of the concepts of nearby nature and its role in nature deficit. Urban planners are implementing park designs that reinforce the important presence of nature in urban life, but their efforts are sometimes thwarted by the realities of the availability of public funding and the accompanying lack of public support. As Miller (2005) noted "If there is to be broad-based public support for biodiversity conservation, the places where people live and work should be designed so as to provide opportunities for meaningful interactions with the natural world" (p. 430).

The field of social-cultural sustainability incorporates issues related to the urban environment, particularly in desert communities where the environment is a crucial component of sustainability. Measures of nature deficit and perceived potential impacts as suggested by various authors (Louv 2008; Hartig et al. 2014; Karieva 2008) provide an innovative perspective on desert community sustainability that remains unexplored to date.

Implications of research into nature deficit are a valuable contribution to the field of park and recreation management. An ongoing concern among park and recreation managers and planners is the declining participation by the U.S. population as a whole at public park and recreation areas, particularly natural areas. Participation in outdoor recreation areas continues to increase, but at a slower rate than the increase of the US population (McLean and Hurd 2015). Using multiple data sets Pergams and Zaradic (2006, 2008), reported a fundamental shift away from nature-based recreation. Declines in participation rates increased from 18 % in 1981 to 25 % in 2008. Evidence indicates that attendance by international visitors at national parks is reaching levels that may soon over take U.S. visitation rates (Meyers 2008). There is additional evidence to suggest that special interest groups or individuals with narrow foci account for an inordinate amount of outdoor recreation usage hence, the decrease in per-capita participation in outdoor recreation areas by the US population may be higher than previously reported by Federal and state agencies.

Pyle (1978) suggested an alienation from nature could result in the extinction of experience (Kellert 2009; Miller 2005). This contributes to the perpetuation of environmental generational amnesia (Kahn 1999, 2002; Wells and Lekies 2006) theorized as a contributing factor in environmental degradation. Nature deficit research significantly informs not only these environmental psychology concepts but the related fields of natural resource management and environmental education since efforts to reduce deficit levels require these fields to collaborate on this human dimension phenomenon.

The field of social-cultural sustainability incorporates issues related to the urban environment, particularly in desert communities where the environment is a crucial component of sustainability. Measures of nature deficit and perceived potential impacts as suggested by various authors (Louv 2008; Pergams and Zaradic 2006, 2008; Karieva 2008; Larson et al. 2011) provide an innovative perspective on a desert community sustainability, which remains unexplored to date.

The examination of nature deficit through the proxy of urban greenspace through time is useful to determine the changing dynamics of the urban area. The fields of urban planning, socio-cultural sustainability, environmental sciences, education, environmental justice, environmental education, resource management, and park and recreation management are but a few of the disciplines involved in examining the concept of nature deficit (Strife and Downy 2009). As Americans move further and further away from nature, of any kind, experience higher levels of nature deficit, and replace nature with videophilia, their perceptions and commitment to nearby nature and a sustainable environment in general will continue to decline at a time when climate change and biodiversity data suggest broad scale concern and care for natural areas is desperately needed. This is especially problematic in urban and suburban areas that have been expanding for many decades. To begin to address these issues, this chapter examines the amount of greenspace and greenspace changes in the greater Las Vegas, Nevada area in 1985 and 2014 using remote sensing data and a vegetation index. It represents the first step in a longer project designed to measure nature deficit in the Las Vegas area.

6.2 Methods

Greenspace was measured in the greater Las Vegas metropolitan area (which includes the cities of Las Vegas, North Las Vegas, and Henderson, as well as portions of unincorporated Clark County, Nevada in the Las Vegas metropolitan area) using remote sensing data and vegetation indices for two dates: 1985 and 2014.

Taking Louv's (2005) references to land development as a corroborating factor into account, Clark County, Nevada is a uniquely qualified area to investigate nature deficit disorder in an arid environment. Encompassing acres of federal open space as well as declining manicured lawns and expanding desert xeriscaping, the Las Vegas metropolitan area was ranked the third fastest growing metro area between 2000 and 2010 and then fell dramatically as a result of the recession, but has since rebounded (retrieved January 13, 2015 from http://www.census.gov/newsroom/releases/archives/population/cb12-55.html). This rapid land development and increasing urbanization of the county juxtaposed by the millions of acres/hectares of federal lands surrounding the urban areas presents an opportunity to examine not only the role of nearby nature (nature in close proximity to urban residents) in nature deficit but the potential for outdoor recreation service providers to assist in the prevention of the formation of nature deficit by facilitating time in nature on public lands.

6.2.1 Remote Sensing Data

This study relies on multispectral data acquired using the Landsat 5 and Landsat 8 (also called the Landsat Data Continuity Mission) satellite systems. Landsat 5 data were acquired in August 1985 and Landsat 8 data were acquired in August 2014. These data were used to calculate the Normalized Difference Vegetation Index (NDVI). Vegetation indices, such as NDVI have been calculated using remote sensing data for many years. These radiometric measures have been used to model Leaf Area Index (LAI), absorbed photosynthetically active radiation (APAR), and the relative abundance and activity of green vegetation (Jensen 2005; Lymburner et al. 2000; Liu and Huete 1995). The ratio of reflected radiance from the NIR and red bands incorporates a significant amount of illumination and topographic variation, and forms a stable indicator of the vigor of green vegetation (Liu and Huete 1995). NDVI has been identified as a useful tool to examine regional, continental, and global photosynthesis (Bonan 1993; Running et al. 1994). NDVI is calculated using the equation (Rouse et al. 1974):

$$(\mathrm{NIR} - \mathrm{Red}) / (\mathrm{NIR} + \mathrm{Red})$$

where NIR is the near infrared reflectance and red is the reflectance. NDVI has been used in phenological studies, in land-cover classification, in global climate change models, and even to predict outbreaks of insect-borne diseases (Justice et al. 1985; Sellers et al. 1994). NDVI has been correlated with biophysical plant canopy properties (Pinty et al. 1993), including its relation to biomass (Pearson and Miller 1972; Tucker 1979; Elvridge and Lyon 1985; Box et al. 1989), LAI (Badhwar et al. 1986; Clevers 1988, 1989; Gholz et al. 1997; Jensen and Binford 2004) primary productivity (Tucker and Sellers 1986), photosynthetic radiation (Asrar et al. 1984; Choudhury 1987), and other ecological parameters (Cihlar et al. 1991).

In this chapter, greenspace was defined by an NDVI value greater than or equal to 0.20. After NDVI was computed, polygons were digitized around both the 1985 and 2014 urban areas to determine the percentage of greenspace in both years.

6.3 Results

NDVI maps were computed for both August 1985 and August 2014 (Figs. 6.1 and 6.2). NDVI values greater than or equal to 0.20 included lawns, golf courses, parks, and other green areas throughout Las Vegas. In 1985, 9.4 % of the Las Vegas urban area was greenspace. In 2014, this changed to 6.9 % of the urban area.

These numbers indicate that the amount of greenspace in the Las Vegas area has generally decreased in the last 30 years – even as the amount of people living in the greater Las Vegas has increased.

Fig. 6.1 NDVI map of the greater Las Vegas area in 1985. *Lighter pixels* represent areas with greater amounts of vegetation

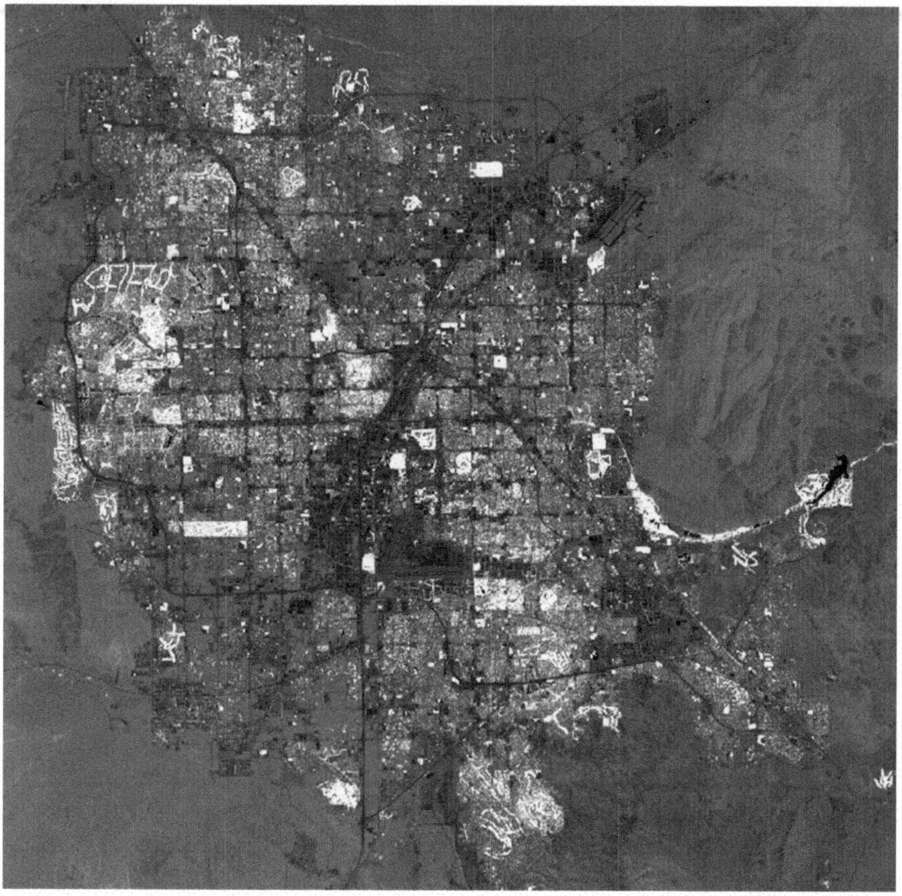

Fig. 6.2 NDVI map of the greater Las Vegas area in 2014. *Lighter pixels* represent areas with greater amounts of vegetation. Note the larger area that Las Vegas now occupies relative to 1985

6.4 Conclusion

The study provides urban planners and park and recreation managers with a method to measure urban greenspace using remote sensing data and methods. These data and methods may assist planners and managers in making decisions when planning facilities and services. Further, although not addressed in this chapter, results from studies like this study and others promise to provide insights into effective strategies for reaching out to underrepresented sociodemographic groups and reducing issues of environmental justice.

The study enhances the ability of geographers, urban planners, and others concerned with the human urban environment to model, explain, and understand urban environments – or establish what Longley (2002, p. 231) calls a 'better' urban geography. Applications of traditional remote sensing methods to urban desert environments remain largely nonexistent, yet according to the Brookings Institute (2008) the urban desert areas in the Intermountain West will experience the largest population growth in the United States. Their "Mega Mountains" report issues a call to Congress to support the development of regional infrastructure in these areas and the ability to accurately utilize remote sensing methods in urban desert areas will play a crucial role in these vital urban planning efforts.

6.5 Discussion

The recognition of Las Vegas as an arid region places it in a unique situation among most metropolitan areas studied in relation to nature and health and the presence of greenspace. Martin (2008) addressed issues of ecoregion and ecosystem resiliency factors that need to come into consideration when attempting to achieve a sustainable urban landscape. Regions such as Las Vegas were initially adapted to fit into a traditional lush landscape with diverse non-native plant taxa similar to cities in a more mesic climate. The sustainability of such landscapes could not be sustained and as early as 1947 changes were made in urban water planning began to reduce traditional lawns with mixed oasis and xeric desert environment plant materials. Hence, as the Las Vegas metropolitan area expanded businesses and homeowners were restricted in the amount of lawn and types of plant materials available.

The result is present in the vegetation change shown between Figs. 6.1 and 6.2. Admittedly, traditional definitions of greenspace in mesic climates likely do not fit an arid ecosystem like the Mojave Desert. From a nature deficit perspective the term greenspace or green needs to be adapted. The presence of greenspace is composed of native and nonnative materials in scattered plots throughout the metropolitan area. The presence of an urban forest is stretched, and many local plant materials can be viewed, but often are not conducive for touching. Residents become aware that greenspace of lawns, parks, and an urban forest are the exception rather than the expected. Research needs to address issues of (1) perceptions of how people perceive nature in a desert environment, (2) how do desert urban landscapes influence individuals attitudes towards the environment, and (3) are outcomes from contact with nature in a desert community different from a mesic community? We suggest that the influence of greenspace upon residents may influence attitudes toward environmental sustainability. While we do not suggest a direct correlation, but based on our observations, we propose a linkage may be present and measurable (Beatley and Newman 2013). Where nature deficit is an absence of contact with nature that must be corrected (Dickinson 2013), sustainability is a life-style adaptation that is influenced by early childhood contact with a "sense of wonder" in the environment (Carson 1956). Many more questions can and should be researched. Making assumptions about the influence of greenspaces in desert regions should be grounded in good research. This chapter opens a door that poses more questions than it answers and ideally encourages further research.

References

Asrar G, Fuchs M, Kanemasu ET, Hatfield JL (1984) Estimating absorbed phtosynthetic radiation and leaf area index from spectral reflectance in wheat. Agron J 76:300–306

Badhwar GD, MacDonald RB, Hall FG, Carnes JG (1986) Spectral characterization of biophysical characteristics in a boreal forest: relationship between Thematic Mapper band reflectance and leaf area index for aspen. IEEE Trans Geosci Remote Sens GE-24:322–326

Beatley T, Newman P (2013) Biophilic cities are sustainable, resilient cities. Sustainability 5(8):3328–3345

Bonan GB (1993) Importance of leaf area index and forest type when estimating photosynthesis in boreal forests. Remote Sens Environ 43:303–314

Box EO, Holben BN, Kalb V (1989) Accuracy of the AVHRR vegetation index as a predictor of biomass, primary productivity and net CO2 flux. Vegetation 80:71–89

Carson R (1956) The sense of wonder. Harper and Row, New York

Choudhury BJ (1987) Relationships between vegetation indices, radiation absorption, and net photosynthesis evaluated by a sensitivity analysis. Remote Sens Environ 22:209–233

Churchman A (2003) Is there a place for children in the city. J Urban Des 8:99–111

Cihlar J, St-Laurent L, Dyer JA (1991) Relation between the normalized difference vegetation index and ecological variables. Remote Sens Environ 35:279–298

Clevers JGPW (1988) The derivation of a simplified reflectance model for the estimation of leaf area index. Remote Sens Environ 25:53–69

Clevers JGPW (1989) The application of a weighted infrared-red vegetation index for estimating leaf area index by correcting for soil moisture. Remote Sens Environ 29:25–37

Dickinson E (2013) The misdiagnosis: rethinking "Nature-deficit Disorder". Environ Commun J Nat Cult 7(3):315–335

Dustin DL, Bricker KS, Schwab KA (2009) People and nature: toward an ecological model of health promotion. Leis Sci 32(1):3–14

Elvridge CD, Lyon RJP (1985) Influence of rock-soil spectral radiation on assessment of green biomass. Remote Sens Environ 17:265–279

Gholz HL, Curran PJ, Kupiec JA, Smith GM (1997) Assessing leaf area and canopy biochemistry of Florida pine plantations using remote sensing. In: Gholz GH, Nakane K, Shimoda H (eds) The use of remote sensing in the modelling of forest productivity. Kluwer, Boston, pp 3–22

Godbey G (2009) Outdoor recreation, health, and wellness. Resources for the Future, Washington, DC, 46p

Hartig T, Mitchell R, De Vries S, Frumkin H (2014) Nature and health. Annu Rev Public Health 35:207–228

Jensen JR (2005) Digital image processing: a remote sensing perspective. Prentice Hall, Upper Saddle River

Jensen RR, Binford MW (2004) Measurement and comparison of leaf area index estimators derived from satellite remote sensing techniques. Int J Remote Sens 25(20):4251–4265

Justice CO, Townshend JR, Holder BN, Tucker CJ (1985) Analysis of the phenology of global vegetation using meteorological satellite data. Int J Remote Sens 6:1271–1318

Kahn P (1999) The human relationship with nature: development and culture. MIT Press, Cambridge

Kahn P (2002) Children's affiliations with nature: structure, development, and the problem of enviromental generational amnesia. In: Kahn P, Kellert S (eds) Children and nature: psychological, sociocultural, and evolutionary investigations. MIT Press, Cambridge

Kaplan R, Austin ME (2004) Out in the country: sprawl and the quest for nearby nature. Landsc Urban Plan 69:235–243

Karieva P (2008) Ominous trends in nature recreation [electronic version]. PNAS 105:2757–2758 from http://www.pnas.org/cgi/doi/10.1073/pnas.0800474105

Kellert S (2009) Biodiversity, quality of life, and evolutionary psychology. In: Sala OE, Meyerson LA, Parmesan C (eds) Biodiversity change and human health. Island Press, Washington, DC

Larson LR, Whiting JW, Green GT (2011) Exploring the influence of outdoor recreation participation on pro-environmental behaviour in a demographically diverse population. Local Environ 16(1):67–86

Liu Q, Huete A (1995) A feedback based modification of the NDVI to minimize canopy background and atmospheric noise. IEEE Trans Geosci Remote Sens 33:457–465

Longley P (2002) Geographical information systems: will developments in urban remote sensing and GIS lead to 'better' urban geography? Prog Hum Geogr 26:231–239

Louv R (2005) Last child in the woods: saving our children from nature deficit disorder, 1st edn. Algonquin Books, Chapel Hill

Louv R (2008) Last child in the woods: saving our children from nature deficit disorder, 2nd edn. Algonquin Books, Chapel Hill

Lymburner L, Beggs PJ, Jacobson CR (2000) Estimation of canopy-average surface-specific leaf area using Landsat TM data. Photogramm Eng Remote Sens 66:183–191

Maller C, Henderson-Wilson C, Pryor A, Prosser L, Moore M (2008) Healthy parks, healthy people: the health benefits of contact with nature in a park context. Deakin University and Parks Victoria, Melbourne, 104p

Martin CA (2008) Landscape sustainability in a Sonoran Desert city. Cities Environ (CATE) 1(2):5

McLean DD, Hurd A (2015) Recreation and leisure in modern society, 10th edn. Jones and Bartlett Publishers, Boston

Meyers AL (2008) Grand Canyon, U.S. seeing more foreign tourists lately. Retrieved 15 Aug 2008 from http://www.suntimes.com/lifestyles/travel/northamerica/1088056,TRA-News-canyon03.article

Miller JR (2005) Biodiversity conservation and the extinction of experience. Trends Ecol Evol 20(8):430–434

Moore RC, Marcus CC (2008) Healthy planet, healthy children: designing nature into the daily spaces of childhood. In: Kellert SR, Heerwagen J, Mador M (eds) Biophilic design: the theory, science, and practice of bring buildings to life. Wiley, Hoboken

Pearson RL, Miller LD (1972) Remote mapping of standing crop biomass for estimation of the productivity of the shortgrass prairie. In: Proceedings of the eighth international symposium on remote sensing of environment. Environmental Research Institute of Michigan, Ann Arbor, pp 1357–1381

Pergams ORW, Zaradic PW (2006) Is love of nature in the US becoming love of electronic media? 16-year downtrend in national park visits explained by watching movies, playing video games, internet use, and oil prices [electronic version]. J Environ Manag 80:387–393

Pergams ORW, Zaradic PW (2008) Evidence for a fundamental and pervasive shift away from nature-based recreation [electronic version]. PNAS 105(7):2295–2300, http://www.pnas.org/cgi/doi/10.1073/pnas.0709893105

Pinty B, Leprieur C, Verstraete MM (1993) Towards a quantitative interpretation vegetation indices – Part 1: biophysical canopy properties and classical indices. Remote Sens Rev 7:127–150

Pyle RM (1978) The extinction of experience. Horticulture 56:64–67

Rouse JW, Haas RH, Schell JA, Deering DW (1974) Monitoring vegetation systems in the Great Plains with ERTS. In: Proceedings, third earth resources technology satellite-1 symposium, NASA SP-351, Greenbelt, pp 3010–3017

Running SW, Justice CO, Solomonson V, Hall D, Barker J, Kaufmann YJ, Strahler AH, Huete AR, Muller JP, Vanderbilt V, Man ZM, Teillet P, Carneggie D (1994) Terrestrial remote sensing science and algorithms planned for EOS/MODIS. Int J Remote Sens 15:3587–3620

Sellers PJ, Tucker CJ, Collatz GJ (1994) A global 10x10 NDVI dataset for climate studies. Part 2- the adjustment of the NDVI and generation f global fields of terrestrial biophysical parameters. Int J Remote Sens 15:3519–3545

Strife S, Downey L (2009) Childhood development and access to nature a new direction for environmental inequality research. Org Environ 22(1):99–122

Taylor AF, Kuo FE, Sullivan WC (2001) Coping with ADD: the surprising connection to green play settings. Environ Behav 33(1):54–77

Tucker CJ (1979) Red and photographic infrared linear combinations for monitoring vegetation. Remote Sens Environ 8:127–150

Tucker CJ, Sellers PJ (1986) Satellite remote sensing of primary production. Int J Remote Sens 7:1395–1416

Turner WR, Nakamura T, Dinetti M (2004) Global urbanization and the separation of humans from nature. Bioscience 54:585–590

Wells NM, Lekies KS (2006) Nature and the life course: pathways from childhood nature experiences to adult environmentalism. Child Youth Environ 16:1–24

Chapter 7
Social Networks, Strategic Doing, and Sustainable Management of Local Food Systems

Neil Reid

Abstract In recent years there has been increasing interest in many communities across the United States in developing local food systems. This has resulted in the creation of Food Policy Councils (FPCs) which have emerged as the entity with primarily responsibility for managing the local food system's growth and development. FPCs are necessary because there is no entity, either in the public or private arena, whose responsibility it is to deal with food and food policy issues. Food Policy Councils and the local food systems whose growth and development they oversee are not hierarchical entities but are loosely formed networks of local stakeholders who have a passion and interest in localizing to the greatest extent possible the system that produces the food that we consume. The success of any network is driven, to a large extent, by the structure and nature of the social relationships of the people that comprise the network. Social Network Analysis (SNA) is a technique that allows network relationships to be analyzed and, among other things, network leaders to be identified and relationship gaps to be discovered which if filled could strengthen the FPC. Healthy and productive relationships enhance the sustainability of both the FPC and the local food system. A significant challenge facing FPCs is delivering value in the form of projects and initiatives that result in tangible progress. This is particular challenging when the members of an FPC are volunteers who have constraints on how much time they can devote to the organization. While FPCs may be tempted to develop a strategic plan as a framework for guiding its actions a more productive avenue might be to utilize a process called Strategic Doing. Strategic Doing emphasizes action over planning and is designed to deliver tangible results more quickly.

Keywords Local food systems • Strategic doing • Social networks • Sustainable management

---

N. Reid (✉)
Department of Geography & Planning and The Urban Affairs Center, University of Toledo, Toledo, OH, USA
e-mail: neil.reid@utoledo.edu

J.D. Gatrell et al. (eds.), *Urban Sustainability: Policy and Praxis*, Geotechnologies and the Environment 14, DOI 10.1007/978-3-319-26218-5_7

7.1 Introduction

> If you have a dream, you can spend a lifetime studying, planning, and getting ready for it. What you should be doing is getting started. Drew Houston (Co-founder and CEO of Dropbox)

As interest in consuming locally-grown food increases more and more communities are recognizing the need to think of growing food not as an isolated activity but as one that is embedded within a larger system. This system comprises a number of connected and inter-related activities including the growing, harvesting, packing, transportation, retailing, consumption, and disposal of food (Fig. 7.1). Within the United States coordination of this system is often the remit of Food Policy Councils (FPCs). The members of an FPC represent various constituent parts of the system and can include farmers, food justice advocates, retailers, chefs, restaurateurs, educators, researchers, government officials, interested citizens, and food processors, wholesalers, and distributors (Hamilton et al. 2012). At the local level most FPCs are entirely independent of government and are structured as self-organizing independent grassroots networks (Harper et al. 2009). The diversity of actors, interests, and goals of FPCs is such that a major challenge they face is that of acting with a unified purpose and accomplishing tasks that result in tangible progress and contribute to the vibrancy and health of a local food system. There is also the challenge of sustainability. There are dozens of Food Policy Councils around the country. However, a visit to the websites of many FPCs suggests that a significant number

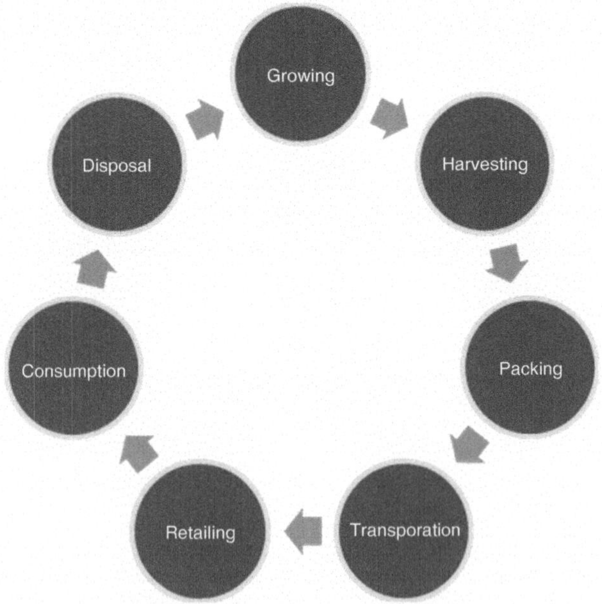

Fig. 7.1 The constituent parts of a local food system

are not particularly vibrant. There appears (at least if the website is any indication) to have been little or no activity for several years. So how does an FPC maintain momentum and achieve sustainability once the initial enthusiasm associated with its creation has died down? While many FPCs have opted to develop a Strategic Plan as its strategic guide there is evidence that such an approach is not always fruitful as plans often sit on shelves unused and gathering dust. An alternative to a Strategic Plan is to engage in a process called Strategic Doing, a new and emerging approach, which emphasizes delivering tangible results quickly.

The remainder of this chapter is divided into six sections. In the next section I outline some of the reasons for increased consumer interest in local food, the suggested benefits of local food, the lack of clarity and agreement as to what constitutes local food, and some of the critiques that have challenged the conventional wisdom of the benefits of local food as promulgated by its advocates. In the third section I examine Food Policy Councils from the perspective of their goals and composition. I also make the point that FPCs are a social network that is embedded within the larger social network that is the local food system itself. I argue that the success of the local food system is to a large extent dependent upon both it and the FPC functioning as a healthy and vibrant social network. In the fourth section I briefly describe some of the key ideas with respect to social networks and their analysis. Key ideas include network density and different measures of the centrality of an individual within a network. The concepts or social capital, trust, local buzz, global pipelines, and absorptive capacity are also mentioned. In the fifth section I describe Strategic Doing and suggest that it as an approach that FPCs might use to ensure that they generate results, build momentum, and move towards sustainability. Strategic Doing is a new and emerging approach that is focused on getting things done and is presented as an alternative to Strategic Planning where progress, when it occurs, tends to be much slower and more labored. In the sixth section I describe my hands-on experience in developing a collaborative network focused around the greenhouse industry in northwest Ohio. This case study example is important because it illustrates the importance of getting things done, some of the strategies that can be used to build a network, and metrics that can be used to gauge its growth and success. In the final section I summarize what I believe are the roles of Social Network Analysis and Strategic Doing in guiding the work of a Food Policy Council as they strive to build a vibrant, healthy, and sustainable food system.

7.2 Local Food

In recent years there has been increased consumer demand for foods that are grown locally (Martinez et al. 2010). This is part of a broader neo-localism movement in which larger numbers of people are demanding products that have a connection with the local community within which they live. Neo-localism has been broadly defined as the "deliberate seeking out of regional lore and local attachment by residents (new and old) as a delayed reaction to the destruction in modern America of traditional bonds to community and family" (Shortridge 1996, 10). The demand for food products with a local connection is apparent in the growing numbers of farmers markets (Cone 2012), community supported agriculture initiatives (Galt et al. 2012), community gardens (National Garden Association 2014), and even craft breweries (Reid et al. 2014). All of these activities and initiatives have prompted Martinez et al. (2010, 1) to refer to the "relocalization" of the food system; a relocalization that stands in sharp contrast to the highly industrialized and globalized system from which most Americans obtain their food (La Trobe and Acott 2000; Godray et al. 2010).

Advocates of local food view it as beneficial for a number of reasons. For example, the University of Vermont Extension identifies ten reasons to buy local food (Grubinger 2010). These are:

- Locally grown food tastes and looks better,
- Local food is better for you,
- Local food preserves genetic diversity,
- Local food is safe,
- Local food supports local families,
- Local food builds community,
- Local food preserves open space,
- Local food keeps taxes down,
- Local food benefits the environment and wildlife,
- Local food is an investment in the future.

While the benefits of local food have been lauded by its advocates there is no universally accepted or legal definition of what constitutes local food (Kremer et al. 2012). The idea of *local* does suggest that a certain geographic proximity exist between producer and consumer, but conceptions of proximity vary considerably. According to the Congress of the United States (2008 Farm Act) local foods are defined as those that are "transported less than 400 miles or within the state in which they are produced" (Johnson et al. 2012). Many consumers disagree with this definition, considering a 400-mile radius to be too generous. A study by the Hartman Group (2008) showed that consumers have a stricter definition of local with 50 % considering "made or produced within one hundred miles" as local while 37 % considered local as "made or produced within my state." Others have suggested that any definition of local food needs to extend beyond the very simple measure of geographic distance and should give consideration to the type of outlet through which the food is marketed, who produced the food, and how it was produced (Thompson et al. 2008; Johnson et al. 2012).

As the theme of this book is urban sustainability it is important to note that a number of communities around the country view local food systems as playing a key role in broader urban sustainability efforts (Morgan 2010). The city of Baltimore defines a sustainable food system as one that "provides healthy food to meet current food needs while maintaining healthy ecosystems that can also provide food for generations to come with minimal negative impact to the environment. A sustainable food system also encourages local production and distribution infrastructures and makes nutritious food available, accessible, and affordable to all" (City of Baltimore 2009, 127). Establishing Baltimore as a leader in sustainable food systems is a key objective within the city's urban sustainability plan and is viewed by the architects of the plan as contributing to broader environmental sustainability.

The benefits of local food have become mantras of the local food movement. However, some authors have questioned the fixation that advocates of local food have on the benefits of 'local'. According to Born and Purcell (2006, 195), "local-scale food systems are equally likely to be just or unjust, sustainable or unsustainable, secure or insecure." They are not inherently more ecologically sustainable, do not automatically result in more social justice, and the product is not necessarily healthier and fresher. "No matter what its scale, the outcomes produced by a food system are contextual: they depend on the actors and agendas that are empowered by the particular social relations in a given food system" (Born and Purcell 2006, 195–196). Born and Purcell (2006) suggest that a critical error often made by local food advocates is that of treating localization of the food system as the outcome rather than as part of a broader strategy to attain some other outcome or set of outcomes – social justice, local economic development, food safety etc. Viewing local food systems through this latter lens inevitably results in the recognition that the creation of a local food system can be characterized as a *wicked problem* (Reid et al. 2012). Wicked problems are those which are notoriously hard to define (e.g., is the problem one of social injustice, food insecurity, or lack of access to healthy food?), tend to require solutions that are context and place specific (what works in Philadelphia might not work in Los Angeles), are never truly solved (there is always room for improvement), whose "solutions" can be difficult to identify (there may be no solution, a single solution, or multiple solutions), and whose "solutions" may result in unintended consequences and may not be easily undone (Rittel and Weber 1973). The inherent wickedness of local food systems means that their creation, sustainability, and criteria by which their success is judged can be challenging and problematic.

7.3 Food Policy Councils

In urban areas of the United States the entity that is most commonly responsible for developing local food systems are Food Policy Councils. As noted by Fox (2010, 2), "the Food Policy Council model has emerged in North America over the last three decades as an attempt to address inadequacies and gaps in food policy and planning. Historically, in most communities there was no single entity responsible for the creation and development of local food systems. In communities where there was an interest in food policy it tended to be "shaped by a disparate array of government departments and agencies without coordination or recognition of the linkages between food-related sectors" (Fox 2010, 2). Meanwhile, planners often ignore food and food policy because they feel that it is a private sector activity that falls outside of their purview (Pothukuchi and Kauffman 2000).

A Food Policy Council comprises a network of individuals (often representing entities) who have a vested interested in the development of a vibrant, healthy, and sustainable local food system. It has been suggested that membership of FPCs should be diverse, not just in terms of their demographic composition but also in terms of the backgrounds, expertise, experience, availability, and capacity for decision making (Hamilton et al. 2012; Pothukuchi et al. 2002). Thus, FPCs often include farmers, food justice advocates, retailers, chefs, restaurateurs, educators, researchers, government officials, educators, interested citizens, and food processors, wholesalers, and distributors (Hamilton et al. 2012). The membership of the average FPC represents 13–14 organizations (National Research Center 2010). This means that a large number of diverse interests are typically represented, each of which may have a different set of priorities, goals, and aspirations for the FPC. A Food Policy Council may be housed within and have a formal relationship with an existing local or state government office or may be an entity that is independent of any government body (Schiff 2008; Harper et al. 2009).

As illustrated in Fig. 7.1 the constituent parts of the food system that an FPC is responsible for coordinating include growing, harvesting, packing, transportation, retailing, consuming, and disposing. While the missions of FPCs differ from place to place the functions they generally perform are:

- serving as forums for discussing food issues,
- understanding the current state of the food system,
- evaluating and influencing food related policy,
- launching and/or supporting programs and services that address local needs
- identifying goals for improvement,
- establishing initiatives to meet goals,
- identifying indicators to monitor the success of food systems initiatives,
- fostering coordination between different sectors in the food system, and
- engaging community members and building community support (Harper et al. 2009; Hamilton et al. 2012).

Given the diversity of individuals, entities, interests, and goals that are typically present within a Food Policy Council and the fact that at the local level FPCs tend to be grassroots organizations it can be particularly informative and useful to view the FPC as a social network. Indeed, Harper et al. (2009, 33) suggest that an FPC is a "network of concerned and engaged individuals and organizations". Not only is the FPC itself a network but the broader local food system that the FPC is responsible for nurturing is also a network. The FPC can, therefore, be thought of a network within a network.

For networks to be effective they have to be understood, nurtured, and managed (Cross and Thomas 2009). One of the key roles that an FPC can play is that of network facilitator. As such an FPC provides a forum where entities and individuals connect, identify what needs to be done, decide what resources are going to be needed, and decide who is going to do what (Schiff 2008). The realization that an FPC is a social network often comprising disparate stakeholders is important for understanding how FPCs (and the local food system whose success depends on its judicious leadership) function and might function more effectively. Viewing the FPC as a network also recognizes the important role played by social relationships in the success and sustainability of a local food system. As suggested by Jarosz (2000, 279) "primary among the influences assuring the viability and vibrancy of regional food networks are the social relations of cooperation and trust that exist among suppliers, producers, workers, brokers, retailers, and consumers who comprise the regional food network." In a similar vein Hamilton et al. (2012) note that successful collaboration is the key to both starting and sustaining a local food system. Furthermore, "examining regional food networks from within and focusing upon the social relations of trust and cooperation among the actors in the network provides a means of identifying obstacles and opportunities within networks" (Jarosz 2000).

7.4 Social Networks and Social Network Analysis

A social network comprises individuals (or organizations) and the relationships that exist between and connect these individuals (or organizations) to each other. In network analysis the individuals (or organizations) are represented by nodes and the relationship by lines or vertices (Carrington and Scott 2011). In Fig. 7.2, for example, the black circles represent individuals. A line connecting two individuals indicates the existence of a relationship between two individuals while the lack of a line indicates that no relationship exists.

Social networks can be analyzed mathematically. This is referred to as Social Network Analysis. Mathematical analysis of any given social network allows the measurement of different characteristics of both the network itself and the individuals within it. While there exists a large scholarly literature on the analysis of social networks there are a few key concepts that are central to a basic understanding of any social network. I will briefly review these in the remainder of this section.

Centrality measures the importance of individuals within a network. Holding a central position within a network often reflects the power, influence, popularity, or prestige that an individual or organization holds (Carrington and Scott 2011). There are a number of different measures of centrality, each of which indicates something slightly different about individuals. Three common centrality measures are *in-degree*, *closeness*, and *betweenness*. In-degree centrality measures the number of times that an individual is sought out by others in the network. For example,

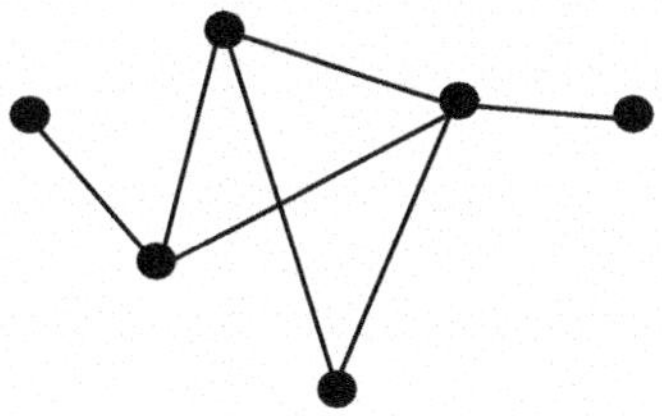

Fig. 7.2 A simple social network

someone whose advice is requested by a large number of individuals would have score high on in-degree centrality (Hanneman and Riddle 2011), indicating that this person is highly respected among network members. Closeness centrality measures the average distance (in terms of number of connections) that an individual is from every other individual within a network. Individuals with high closeness scores occupy a position where they can more easily reach all the other members of a network. This access potentially confers an advantage upon these individuals when it comes to transmitting information within a network. Finally, betweeness (the degree to which someone stands between others within a network) indicates the potential for an individual to "facilitate, impede or bias the transmission of messages" (Freeman 1977, 36). Individuals with high betweeness scores are referred to by a number of different names including *gatekeepers* or *brokers* as they connect individuals who would not be able to otherwise reach each other but for the connection that the gatekeeper/broker provides (Hawe et al. 2004; Long et al. 2013).

In addition to the positions that individuals hold within a network there are also characteristics of a network as a whole that can be important. One of the key characteristics is network density which is simply the percentage of actual relationships to potential relationships. If everyone in a network has a relationship with everyone else within the network then the network density is 100 % (Carrington and Scott 2011). With respect to network efficiency there are advantages of highly dense networks in which a large percentage of network members have a relationship with each other. In a dense network information can flow quickly through the network and trust can be built up more effectively (Freeman 2011). If there is sufficient trust "binding" occurs and the probability that that network members can act collectively as a cohesive unit is increased (Marin and Wellman 2011). Sufficient levels of trust are critical to the success of any network and can be linked to the idea of social capital. Cohen and Prusak (2001, 4) define social capital as "the stock of active connections among people: the trust, mutual understanding, and shared values and behaviors that bind the members of human networks and communities and make cooperative action possible". Trust is something that is built-up over time (Lorenz 1999). The most effective way to build trust within a network is to get network members to actually engage in collaborative projects (Reid and Carroll 2008; Lorenz 1999). In a study of the machine building industry in Lyon, France, for example, Lorenz (1999, 309) noted that "firms should start by making small commitments to each other and progressively increase their commitments depending on the quality of the exchange . . . in short, trust was built up through a learning process. Small risks were followed by larger ones, contingent on the success of the cooperation." Starting with a low-risk project that had a relatively high probability of success and which demonstrated the value of collaboration was the approach successfully adopted by Reid and Carroll (2008) in getting greenhouse growers in northwest Ohio to work together to address common challenges.

In low-density networks, where there is a paucity of connections, it is possible to strategically facilitate the creation of relationships through the process of network weaving (Krebs and Holley 2002). Economic geographers refer to network weaving as creating "local buzz" (Bathelt et al. 2004). Local buzz is defined as "the information and communication ecology created by face-to-face contacts, co-presence and co-location of people and firms within the same industry and place or region" (Bathelt et al. 2004, 38). In the language of social network analysis the creation of a dense network results in "strong ties" (Borgatti and Lopez-Kidwell 2011) – many members of a network know each other and have many of the same relationships in common. Network weaving is often most effective when it is the responsibility of a small number of individuals (network weavers). In the absence of network weaving there is a tendency for relationship patterns to become fossilized. New relationships that depend on serendipity or spontaneity tend to develop very slowly or may not occur at all (Krebs and Holley 2002).

In addition to strategically building relationships within a network it is important for the network weavers to look beyond the network. A danger associated with dense fossilized networks is that much of the information and many of the ideas flowing through the network become quickly redundant with the result that the creation of new information, ideas, and solutions tends not to occur. To combat the challenge of information redundancy it is important that a network connect with other networks. Economic geographers refer to these external connections as "global pipelines" (Bathelt et al. 2004). Global pipelines provide a network with access to "bodies of knowledge residing elsewhere" (Bathelt et al. 2004, 33) and become critical sources of new information and ideas. Access to other networks with new ideas or information requires that at least one individual within a network have a relationship with someone in another network. This inter-network relationship is an example of what Granovetter (1973) refers to as a "weak tie" and individuals who have such connections are referred by a number of terms including *broker, bridger, and gatekeeper* (Hoppe and Reinelt 2010; Long et al. 2013). For example, Fig. 7.3 shows three networks and the relationships that exist both within and between them. In Network A there is an individual who can connect the net-

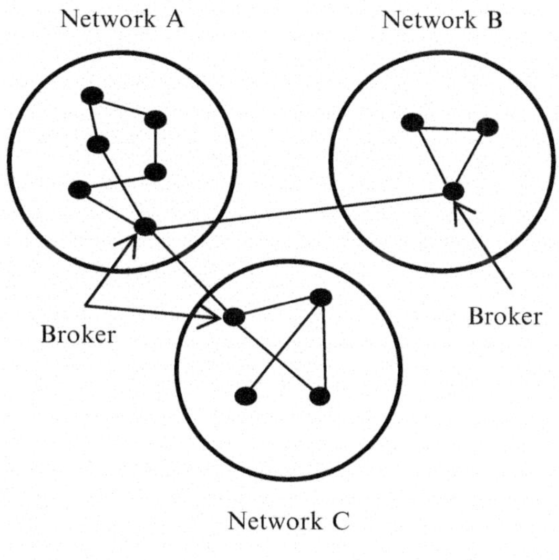

Fig. 7.3 Networks and brokers

work with Networks B and C because of a relationship that he/she has with individuals in each of these networks. Likewise both Networks B and C have individuals who function as brokers to Network A. However, there is no individual in Networks B and C who can serve as a broker between those two networks.

For a network to take advantage of global pipelines its members must be open to new ideas and must have the desire and ability to adapt and use them to beneficial ends. The extent to which a network is able to do this is referred to as its *absorptive capacity* (Escribanoa et al. 2009; Camisón and Forés 2010). Thus even though new ideas may be made available to a network its members may not have the desire and/or ability to utilize these ideas to the network's advantage if it has low absorptive capacity.[1]

Some of the questions that a social network analysis can help answer are (Krebs and Holley 2002; McGregor 2006):

- Are the right connections in place? Are any key connections missing?
- Who are playing leadership roles in the community? Who is not, but should be?
- Who are the experts in process, planning and practice?
- Who are the mentors that others seek out for advice?
- Who are the innovators? Are ideas shared and acted upon?
- Are collaborative alliances forming between local stakeholders?

7.5 Strategic Doing

To be successful a Food Policy Council needs to get things done. Before engaging in projects or initiatives, however, it is not unusual for an FPC to undertake the process of Strategic Planning (Agnew: Beck Consulting 2012; Burgan and Winne 2012). In reality, however, the creation of a strategic plan may not be the most effective way for a Food Policy Council to develop and nurture a local food system. For a variety of reasons, Strategic Planning often fails to deliver on its promises. Reid (1989, 556) describes strategic planning as "a sterile process, one which is oriented merely to the production of documents, the existence of which frequently fails to result in any meaningful change in the behavior of organizations." Strategic planning often disappoints because the energy and commitment that is required to integrate the plan into the day-to-day operations of an organization is frequently lacking with the result that "plans are often committed to drawers to gather dust until the next planning cycle" (Reid 1989, 563). An alternative to Strategic Planning does exist, however, in the form of Strategic Doing. Strategic Doing "is a new strategy discipline specifically designed for open, loosely-connected networks. Unlike strategic planning that was designed primarily to guide strategic activity in hierarchical

[1] There exists a plethora of software packages that can be used conduct a social network analysis. A good review of some of these can be found in Huisman and van Duijn (2011). A more comprehensive list of software packages can be found at http://en.wikipedia.org/wiki/Social_network_analysis_software

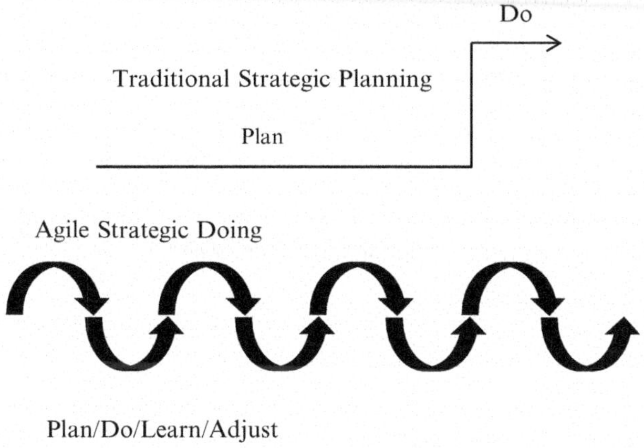

Fig. 7.4 Strategic planning versus strategic doing (Source: Morrison 2013)

organizations, Strategic Doing is designed for situations in which nobody can tell anybody else what to do." (Center for Regional Development, Purdue University 2014). It is a strategy that "enables people to form action-oriented collaborations quickly, move them toward measurable outcomes, and make adjustments along the way" (Morrison and Hutchison 2014) and as such would seem to be particularly appropriate tool for Food Policy Councils to use in their development of local food systems (Fig. 7.4).

Strategic Doing (Fig. 7.5) works by guiding a network towards a series of strategic actions by focusing on the following four questions[2]:

- What could we do?
- What should we do?
- What will we do?
- What is our 30/30?

"What could we do?" – One of the characteristics of Strategic Doing is that it builds upon existing assets (both tangible and intangible) that exist within a network. Assets include expertise, facilities, money, time etc. Network members should begin by identifying network assets (especially those that might be available to the network) while simultaneously considering ways in which different assets could be combined to solve a particular problem or set of challenges. Through this process of guided discussion opportunities for collaborative problem solving are identified.

What should we do? – Having identified the opportunities for collaborative action the next stage is to identify the most promising opportunities. These opportunities will become projects upon which network members will collaboratively work on. Not all the opportunities identified in the "What could we do?" stage

[2] In this chapter Strategic Doing is only very briefly outlined. More detail about how to implement a Strategic Doing approach can be found at https://www.pcrd.purdue.edu/signature-programs/strategic-doing.php and in Morrison (2012).

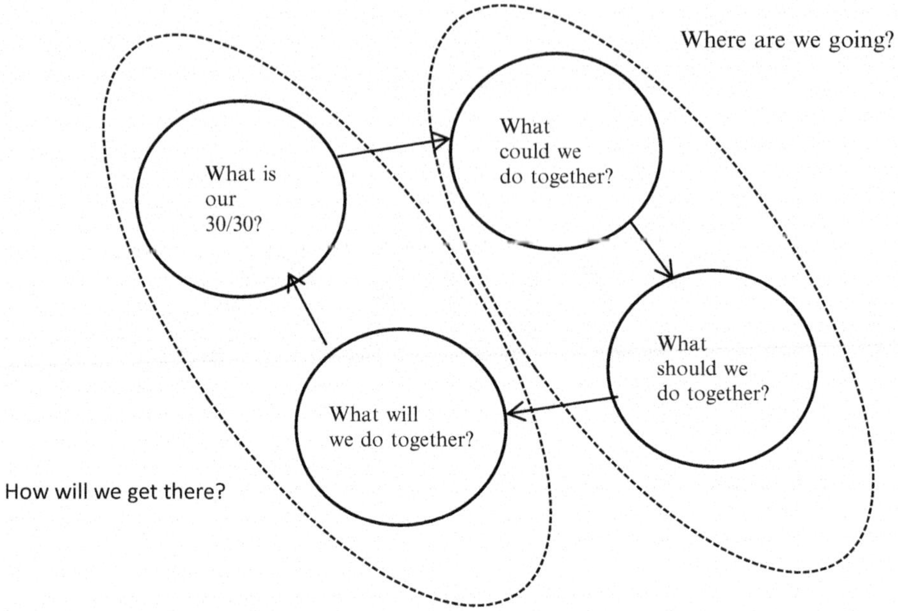

Fig. 7.5 The strategic doing cycle (Source: Morrison 2013)

will result in actionable projects. Indeed, in order to get started the network may identify just one project that it wishes to undertake. Additional projects can be identified at a later date. The identified project(s) must be given concise and measurable outcomes. It is often best to start with a project or projects that are relatively easy to complete. At the same time, however, the initial project(s) should result in outcomes and achievements that are meaningful, provide demonstrable value, and fit within the broader scope of what the network is trying to achieve (e.g., a more sustainable local food system).

What will we do? – When the most promising opportunity or opportunities have been identified and articulated as projects the next step is to develop an action plan that identifies what steps need to be taken towards operationalizing and moving the projects forward. As part of the action plan individuals should volunteer to undertake specific tasks. Each task should have an agreed upon date for completion, preferably by the time of the next meeting (see below). The idea here is that everyone in the network takes on the responsibility for at least one task, even if that task is relatively minor (e.g., sending out some e-mails). This ensures that everyone participates and that the work is shared among network members. This last point may be particularly important in a network where everybody is busy and has limited time to take on additional responsibilities. The old adage many hands make light work comes to mind.

What is our 30/30? – In a Strategic Doing approach the members of the network come together every 30 days. In doing so they answer two 30/30 questions. First, what did you do during the last 30 days? Second, what are you going to do during the next 30 days? Coming together every 30 days allows network members to report on progress, identify adjustments that need to be made, and agree upon next steps. Ongoing meetings, every 30 days, allows for learning, and adjustment to occur and recognizes the reality that many of the problems we deal with are *wicked* problems and that strategy is an ongoing process that are never solved in the conventional sense but always remain a work in progress.

While a relatively new approach Strategic Doing has been successfully applied in a number of places and contexts. These include accelerating innovation in workforce development in north central Indiana, rebuilding neighborhoods in Flint, Michigan, and launching clean energy, aerospace, and water technology clusters in Brevard County, Florida, Rockford, Illinois, and Milwaukee, Wisconsin respectively (Morisson 2013).

7.6 The Northwest Ohio Greenhouse Industry: A Model of Collaboration and Network Building

An understanding and appreciation of social networks and the importance of getting things done can be helpful to a Food Policy Council as it works to build a vibrant, healthy, and sustainable food system. To illustrate this I will document my own experience in working with the northwest Ohio greenhouse industry between 2004 and 2011.[3] I chose to do so as it provides an actual example of a social network in action. As this experience has been documented extensively elsewhere (Reid and Carroll 2006; Gatrell et al. 2012) I will describe it fairly briefly here. In 2004 the University of Toledo received funding from the U.S. Department of Agriculture (USDA) with the remit that the monies would be used to help the northwest Ohio greenhouse industry identify and address its major competitive challenges. These challenges were numerous in nature and included international competition, high energy costs, ageing infrastructure and technology, and a weak market presence. The industry comprised approximately 80 greenhouses, all of whom were family-owned and focused on the production of floriculture products. To help the growers address these challenges we formed an advisory group. While its size and composition changed over time the initial advisory group comprised 15 members representing greenhouse growers, the agricultural extension service, academia, government, and economic development agencies. This advisory group was analogous to a Food Policy Council in that their charge was to identify projects and initiatives upon which the region's growers could collaborate to enhance the industry's economic

[3] As a result of funding provided by the USDA the University of Toledo was formally involved with this project between 2004 and 2011. Despite formal university involvement ending in 2011 collaboration between the growers continues. Information about the project can be found at www.maumeevalleygrowers.com. Other university partners including Bowling Green State University, Indiana State University, and the Ohio State University also partnered with the University of Toledo during various stages of this project.

competitiveness. The inclusion of non-industry members on the advisory group was important because it was recognized that many of the initiatives would require partners outside of the greenhouse industry. In addition to the advisory group there were two paid staff members, a Project Manager and a Champion. Their charge was to function as liaisons between the advisory group and the broader industry and non-industry community, to facilitate implementation of collaborative projects, and to work as network weavers. The reason that two individuals (and not one) were chosen to perform these tasks is that each brought distinctive and non-overlapping competencies to the project. The Project Manager was an industry outsider who had previous experience as a small business owner, an economic development official with a local municipality, and a project manager on a similar university-led project. These varied past experiences meant that he had developed meaningful relationships with a broad range of individuals within the northwest Ohio community. The greenhouse industry was relatively inward looking and having an individual who had a broad set of connections and relationships throughout the region would prove critical to introducing growers to both new ideas and non-industry partners. In contrast, the Champion was an industry insider, a retired agricultural extension agent who knew and understood the industry and was universally respected by the growers.

Starting in October 2004 the advisory group met once a month to identify common challenges and collaborative solutions to these challenges. It was then the task of the Project Manager and Champion to work with the growers to implement identified solutions. It was important that the group minimize the amount of time spent talking about and maximize the amount of time spent doing projects. The growers were extremely busy individuals who did not have time to spend in meetings that did not lead to actionable initiatives. While the advisory group was unaware of the concept of Strategic Doing at the time I realized a number of years later that the approach that the group took was very similar to that advocated by Strategic Doing. We identified collaborative opportunities strategically, but quickly, and moved to action as soon as we could. Monthly meetings were used to update the group on progress with respect to ongoing initiatives, identify what needed to be done (and by whom) within the next 30 days, and to discuss new potential collaborative opportunities.

Over the years the growers participated in a number of collaborative initiatives that enhanced their economic competitiveness. An example of a project was branding and marketing. This was the first collaborative initiative undertaken by the growers and was a response to the challenges of international competition and a stagnant market. The growers agreed that their market presence could be enhanced by the development of a brand identity for the industry and collaborative marketing efforts. The result was the creation of the Maumee Valley Growers brand and the implementation of ongoing collaborative marketing efforts.

It is important to note that not all growers participated in collaborative marketing or any of the other collaborative initiatives (e.g., energy purchases, recycling, and philanthropic giving). Indeed each initiative brought together a different network of growers and non-industry collaborators. The list of non-industry collaborators has been particularly impressive. Since 2004 the growers have worked with several dozen non-industry partners including the Catholic Diocese of Toledo, Toledo Area Regional Transit Authority, the City of Toledo, and Toledo Botanical Gardens. One of the key tasks of the Project Manager was to use his broad base of connections within the local community to identify and introduce the growers to the potential project partners. In other words the Project Manager played a bridging role between the growers and the broader community. A 2007 Social Network Analysis showed the geography of the collaborative relationships that growers were engaged in (Fig. 7.6). These were primarily inside northwest Ohio although a few did extend beyond the local region.

While the success of the northwest Ohio greenhouse project has been documented elsewhere (Reid and Smith 2012) it would be informative to highlight a few key changes that have occurred as a result of the growers coming together and working together on collaborative projects. A key objective of the university researchers that led this project was that of sustainability. We were aware of the possibility that

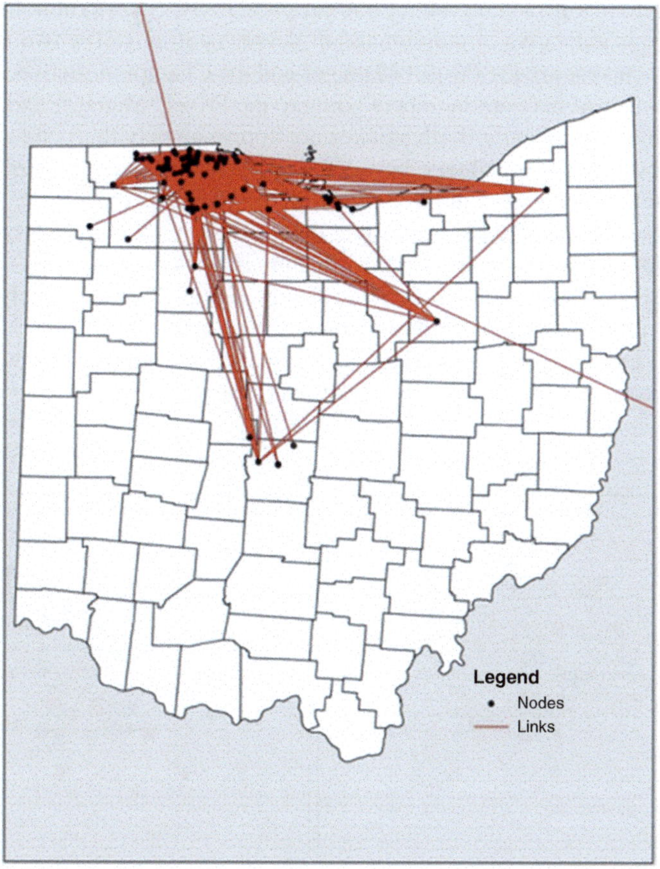

Fig. 7.6 Northwest Ohio greenhouse industry collaboration network, 2007

when federal funding for this project was terminated (as it was) that the growers would stop meeting and collaborative initiatives would cease. Key strategies in guarding against this possibility were to (a) demonstrate the value that collaboration can bring and (b) to build and strengthen the social network of relationships between growers. With respect to demonstrating the value of collaboration 55 % of growers (in a 2009 survey) said that participation in the project had helped them to attract new customers and/or serve existing customers better, while 60 % indicated that they were optimistic about the future of their business (up from 27 % in 2004) (Reid and Smith 2012). With regard to strengthening relationships between growers nearly 80 % of growers said that they have more interaction with other growers as a result of participating in this project (Reid and Smith 2012). Figure 7.7a, b which show the results of Social Network Analyses that were conducted in 2007 and 2009 are also informative in this regard. In both diagrams the red squares represent individual growers and the blue squares represent the Project Manager, Champion, and university researchers. The size of each square is a measure of each individual's centrality within the network. The larger the square the more people within the network that go to that person for advice and support. To use SNA terminology the size of the square is indicative of individuals' in-degree centrality. The two largest blue squares in each year are the Project Manager and the Champion. Between 2007 and 2009 the number of network members going to the Project Manager and Champion did not change significantly (both squares are approximately the same size in each year). However, what did change between 2007 and 2009 was the size of the red squares. The in-degree centrality of many of the growers (red squares) increased between these 2 years. This indicates that the growers were increasingly turning to each other for advice and support. This is a sign of a healthy network.

Figure 7.8 provides another example of how Social Network Analysis can be used in providing valuable information about the members of a network. Using the *betweenness* and *in-degree* measure of centrality we were able to show which members of the network in northwest Ohio had both (a) strong connections to individuals

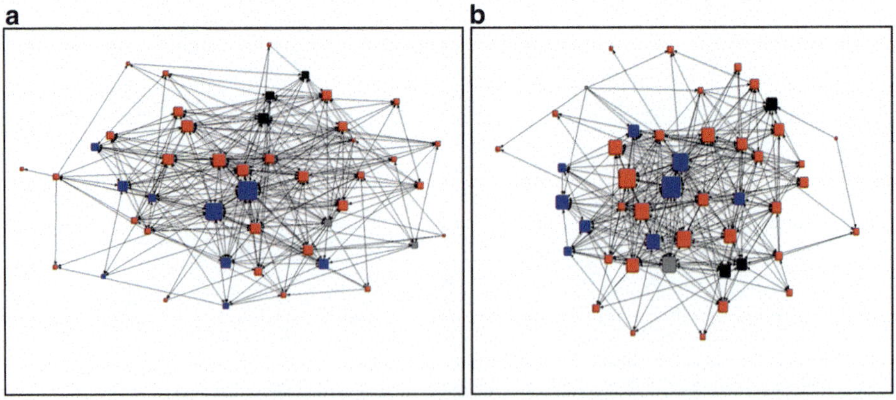

Fig. 7.7 (**a, b**) Social network analysis, northwest Ohio greenhouse industry, 2007 and 2009

Fig. 7.8 Gatekeepers in the northwest Ohio greenhouse industry, 2007

outside of the region and (b) strong connections to individuals within northwest Ohio. Not only do these individuals have potential access to ideas and information from outside of the region but they also are sufficiently respected by their peers in the region that when introducing new ideas and information they are likely to be listened to.

Another indicator that the network had matured in a healthy direction was that 60 % of the growers indicated that they felt that they had enhanced access to university researchers as a result of their participation in the project. Where appropriate, and when invited to do so, universities can play a valuable role in helping organizations, industries, and regions in their development efforts (Gunasekara 2006; Benneworth and Hospers 2007a, b). They can help build local buzz as well as establish global pipelines. Through the process of convening monthly meetings in which growers focused on collaboratively addressing common challenges the University of Toledo facilitated the enhancement of healthy local buzz. The university faculty who were involved in the project provided a source of new ideas to the growers and conducted research on topics (e.g., on customer demographics) that the growers identified as useful.

While the growers did have some connections outside of northwest Ohio (Fig. 7.6) the university researchers recognized the need for growers to be exposed to ideas from even further afield. To do this the university (in 2010 and 2011) led two study tours to the Flanders region of Belgium (Miller et al. 2015). Growers, university researchers, the Project Manager, and Champion participated in these study tours. Flanders was chosen because it is one of Europe's leading greenhouse regions. The trip to Belgium was facilitated by the fact that one of the members of the northwest Ohio greenhouse network had a family member who was working within the greenhouse industry in Flanders. In social network terminology this individual was a bridger and through his family connection was able to connect the northwest Ohio greenhouse network with the industry in Flanders. The premise behind the visits to Flanders was to identify practices that could be introduced to the northwest Ohio industry. This would be a test of the absorptive capacity of the northwest Ohio greenhouse industry. Also, some northwest Ohio growers were also interested in expanding the amount of their greenhouse space devoted to vegetables and Flanders was one of Europe's leading greenhouse vegetable growing regions. As a result of these two visits a number of growers did introduce some practices that they saw in Flanders. These included experimenting with alternative lighting methods, introducing bumblebees for the purpose of natural pollination, trying different methods of plant pruning, using low-cost plastic stem stabilizers for maximum yield support, packaging plants more attractively to increase sales, and using higher quality plant media for growing.

7.7 Discussion

The case study of the northwest Ohio greenhouse industry can provide a number of lessons for Food Policy Councils that are working on the development of a local food system. While these lessons can perhaps be most effectively applied to a newly formed Food Policy Council they can still bring value to FPCs that have been in existence for a number of years. To some extent the lessons are fairly straightforward and simple. First, focus on doing projects rather than engaging in a broader strategic planning effort. If an FPC has already completed a strategic plan or feels compelled to do one then the Strategic Doing approach can be used to operationalize the Strategic Plan. It is important to focus on what can be achieved with respect to strengthening the local food system and not spending time worrying over what cannot be done. In other words focus on the doable. Start the process of collaboration with identifying projects that are easy to do. Early success builds confidence, increases trust among collaborators, and shows that value and benefits of collaboration. A series of small achievements, considered in combination, can add up to significant progress and can lay the groundwork for continued and more complex and meaningful collaborations in the future. In the case of the northwest Ohio greenhouse industry project there were paid staff people who were dedicated to making sure that things got done. This was definitely beneficial to the project. However, the Strategic Doing approach does suggest that with a sufficient number of dedicated volunteers who are willing to share the workload significant progress can be made and gains achieved. Understanding that the Food Policy Council and the local food system that it is trying to nurture are both networks are critical to the success of both. Unlike hierarchical organizations (e.g., a company) where someone is in charge and can tell subordinates what to do the success of networks depends on individuals and entities voluntarily working together to achieve a common purpose. As such it is important to understand the structure of the network, the relationships within that network, and the position that individuals hold and the roles that they play. Social Network Analysis can contribute to this understanding. The benefits of SNA are outlined above and an FPC can benefit from undertaking such an exercise. Among other things an SNA allows an FPC to see who has relationships with whom, the strength of these relationships, and which individuals appear to be the most respected and trusted individuals in a network. A noted by Morrison (2012, 157) "everyone has a network, and when we choose to collaborate, we bring this network into the room". Just as importantly an SNA allows an FPC to see who is not connected and does not have a relationship with each other. If connecting non-connected people can potentially benefit the development of the local food system then there is a need for some network weaving to occur so that these non-connected individuals can become connected. Food Policy Councils must also be aware of the importance of bringing new ideas to the table and know which individuals have which connections to other individuals or entities outside of the immediate network can be identified through Social Network Analysis. While valuable external connections may exist with potential partners outside of the region (even in other countries) it is also possible that an as yet unknown and untapped resource may exist just down the street. Food Policy Councils may also benefit from reaching out to institutions of higher education within their community. Increasing numbers of universities are including community engagement as a core activity within their missions (Benneworth et al. 2013). Universities are a treasure chest of expertise and identifying faculty members who are willing to contribute their particular expertise to the work of a Food Policy Council and the development of a local food system can bring significant benefits.

Strong networks and a track record of success in getting things done are vital to the longevity and sustainability of a Food Policy Council and a local food system. The magnitude and scale of early successes are less important than the fact that members of the Food Policy Council are able to work together (with other partners) to achieve a tangible outcome that results in progress towards the larger goal of a vibrant and healthy local food system. Larger, more significant, successes will come at a later date as relationships evolve, trust is enhanced, new capacities are discovered, and progress is made.

References

Agnew:Beck Consulting (2012) Alaska food policy council strategic plan, Agnew:Beck Consulting, Anchorage. http://www.jhsph.edu/research/centers-and-institutes/johns-hopkins-center-for-a-livable-future/_pdf/projects/FPN/startegic_plans/AFPC_2012_Strategic_Plan_FINAL.pdf. Last accessed 16 Apr 2015

Bathelt H, Malmberg A, Maskell P (2004) Clusters and knowledge: local buzz, global pipelines and the process of knowledge creation. Prog Hum Geogr 28(1):31–56

Benneworth P, Hospers G-H (2007a) Urban competitiveness in the knowledge economy: Universities as new planning animateurs. Prog Plan 67(2):105–197

Benneworth P, Hospers G-H (2007b) The new economic geography of old industrial regions: Universities as global-local pipelines. Environ Plan C Gov Policy 25(6):779–802

Benneworth P, Charles D, Hodgson C, Humphrey L (2013) The relationship of community engagement with universities' core missions. In: Benneworth P (ed) University engagement with socially excluded communities. Springer, Dordrecht, pp 85–101

Boragtti SP, Lopez-Kidwell V (2011) Network theory. In: Scott J, Harrington PJ (eds) The SAGE book of social network analysis. Sage, London, pp 40–54

Born B, Purcell M (2006) Avoiding the local trap: scale and food systems in planning research. J Plan Educ Res 26(2):195–207

Burgan M, Winne M (2012) Doing food policy councils right: a guide to development and action. Mark Winne Associates, Santa Fe

Camisón C, Forés B (2010) Knowledge absorptive capacity: new insights for its conceptualization and measurement. J Bus Res 63(7):707–715

Carrington PJ, Scott J (2011) Introduction. In: Scott J, Harrington PJ (eds) The SAGE book of social network analysis. Sage Publications, London, pp 1–8

Center for Regional Development, Purdue University (2014) Strategic doing https://www.pcrd.purdue.edu/signature-programs/strategic-doing.php. Last accessed 16 Apr 2015

City of Baltimore (2009) The Baltimore sustainability plan. http://www.baltimoresustainability.org/sites/baltimoresustainability.org/files/Baltimore%20Sustainability%20Plan%20FINAL.pdf. Last accessed 16 Apr 2015

Cohen D, Pusak L (2001) In good company: how social capital makes organizations work. Harvard Business School Press, Boston

Cone T (2012) Number of US famers markets surges. USA Today, 3 August. http://usatoday30.usatoday.com/news/nation/story/2012-08-03/farmers-markets-increasing/56727330/1. Last accessed 16 Apr 2015

Cross R, Thomas RT (2009) Driving results through social networks. Jossey-Bass, San Francisco

Escribanoa A, Fosfuri A, Tribó JA (2009) Managing external knowledge flows: the moderating role of absorptive capacity. Res Policy 38(1):96–105

Fox C (2010) Food policy councils: innovations in democratic governance for a sustainable and equitable food system. UCLA Urban Planning Department. http://www.jhsph.edu/research/centers-and-institutes/johns-hopkins-center-for-a-livable-uture/_pdf/projects/FPN/startegic_plans/Food%20Policy%20Councils%20Innovations.pdf. Last accessed 16 Apr 2015

Freeman LC (1977) A set of measures of centrality based on betweenness. Sociometry 40(1):35–41

Freeman LC (2011) The development of social network analysis – with an emphasis on recent events. In: Scott J, Harrington PJ (eds) The SAGE book of social network analysis. Sage, London, pp 26–39

Galt RE, O'Sullivan L, Beckett J, Hiner CC (2012) Community supported agriculture is thriving in the Central Valley. Calif Agric 66(1):8–14

Gatrell JD, Thakur R, Reid N, Smith BW (2012) Clusters and "listening": situating local economic development. J Appl Res Econ Dev 7(1):14–25

Godfray HCJ, Crute IR, Haddad L, Lawrence D, Muir JF, Nisbett N, Pretty J, Robinson S, Toulmin C, Whiteley R (2010) The future of the global food system. Phil Trans R Soc 365:2769–2777

Granovetter MS (1973) The strength of weak ties. Am J Sociol 78(6):1360–1380
Grubinger V (2010) Ten reasons to buy local food. The University of Vermont Extension, Burlington, http://www.uvm.edu/vtvegandberry/factsheets/buylocal.html. Last accessed 16 Apr 2015
Gunasekara C (2006) Reframing the role of universities in the development of regional innovation systems. J Technol Tran 31(1):101–113
Hamilton J, Kohut H, Molina J, Overton M (2012) Collaborative planning for local food systems: municipal priorities in action. Tufts University Department of Urban and Environmental Policy and Planning, Medford, http://ase.tufts.edu/uep/degrees/field_project_reports/2012/Team_5_Final_Report_2012.pdf. Last accessed 16 Apr 2015
Hanneman RA, Riddle M (2011) Concepts and measures for basic network analysis. In: Scott J, Harrington PJ (eds) The SAGE book of social network analysis. Sage, London, pp 340–369
Harper A, Shattuck A, Holt-Giménez E, Alkon A, Lambrick F (2009) Food policy councils: lessons learned. Institute for Food and Development Policy, Oakland
Hartman Group (2008) Consumer understanding of buying local. Hartman Group, Bellvue, http://hartbeat.hartman-group.com/article/220/Consumer-Understanding-of-Buying-Local. Last accessed 16 Apr 2015
Hawe P, Webster C, Shiell A (2004) A glossary of terms for navigating the field of social network analysis. J Epidemiol Community Health 58(12):971–975
Hoppe B, Reinelt C (2010) Social network analysis and the evaluation of leadership networks. Leadersh Q 21(4):600–619
Huisman M, van Duijn AJ (2011) A reader's guide to SNA software. In: Scott J, Harrington PJ (eds) The SAGE book of social network analysis. Sage, London, pp 578–600
Jarosz L (2000) Understanding agri-food networks as social relations. Agric Hum Values 17:279–283
Johnson R, Cowan T, Aussenberg RA (2012) The role of local food systems in U.S. farm policy. Congressional research service report, no. R42155, Washington, DC
Krebs V, Holley J (2002) Building smart communities through network weaving. http://www.orgnet.com/BuildingNetworks.pdf. Last accessed 16 Apr 2015
Kremer P, DeLiberty TL, Schreuder Y (2012) Defining local food systems. In: Reid N, Gatrell JD, Ross PS (eds) Local food systems in old industrial regions: concepts, spatial context, and local practices. Ashgate, Aldershot, pp 147–168
La Trobe L, Acott TG (2000) Localising the global food system. Int J Sustain Dev World Ecol 7(4):309–320
Long JC, Cunningham FC, Braithwaite J (2013) Bridges, brokers and boundary spanners in collaborative networks: a systematic review. BMC Health Serv Res 13(4):2–13
Lorenz E (1999) Trust, contract and economic cooperation. Camb J Econ 23(3):301–315
Marin A, Wellman B (2011) Social network analysis: an introduction. In: Scott J, Harrington PJ (eds) The SAGE book of social network analysis. Sage, London, pp 11–25
Martinez S, Hand M, Da Pra M, Pollack S, Ralston K, Smith T, Vogel S, Clark S, Lohr L, Low S, Newman C (2010) Local food systems: concepts, impacts, and issues. United States Department of Agriculture, Economic Research Service Economic Research Report, Number 97, May 2010
McGregor J (2006) Game plan: first find the leaders. BloombergBusinessWeek Magazine, 20 August. http://www.businessweek.com/stories/2006-08-20/game-plan-first-find-the-leaders. Last accessed 16 Apr 2015
Miller D, Calzonetti F, Reid N, Gatrell JD (2015) Breaking the lock-in to past greenhouse practices by exposing growers to aspirational regions: the comparison of the Northwest Ohio greenhouse industry to Flanders, Belgium. In: Dannenberg P, Kulke E (eds) Economic development in rural areas. Ashgate, Aldershot, pp 51–68
Morgan K (2010) Feeding the city: the challenge of urban food planning. Int Plan Stud 14(4):341–348

Morrison E (2012) Strategic doing for community development. In: Walzer N, Hamm GF (eds) Community visioning programs. Routledge Publishing Group, New York, pp 156–177

Morrison E (2013) "Strategic doing": a new discipline for developing and implementing strategy within loose regional networks. Presented to the Australia-New Zealand Regional Science Association International Annual Conference, 3–6 December, 2013, University of Southern Queensland, Hervey Bay, Queensland, Australia. https://www.pcrd.purdue.edu/files/media/Strategic-Doing-A-New-Discipline-for-Developing-and-Implementing-Strategy-within--Loose-Regional-Networks.pdf. Last accessed 16 Apr 2015

Morrison E, Hutchison S (2014) Accelerating civic innovation through "Strategic Doing". Stanford Social Innovation Review, 20 June. http://www.ssireview.org/blog/entry/accelerating_civic_innovation_through_strategic_doing. Last accessed 16 Apr 2015

National Gardening Association (2014) Garden to table: a 5-year look at food gardening in America. National Gardening Association, Williston

National Research Center (2010) Community food projects: indicators of success executive summary. Community Food Security Coalition, National Research Center, Portland

Pothukuchi K, Kauffman JL (2000) The food system: a stranger to the planning field. J Am Plann Assoc 66(2):113–124

Pothukuchi K, Joseph H, Burton H, Fisher A (2002) What's cooking in your food system? A guide to community food assessment. Community Food Security Coalition, Venice

Reid DM (1989) Operationalizing strategic planning. Strateg Manag J 10(6):553–567

Reid N, Carroll MC (2006) Collaborating to compete: the case of the northwest Ohio greenhouse industry. In: Gatrell JD, Reid N (eds) Enterprising worlds: a geographic perspective on economics, environments, and ethics. Springer, Dordrecht, pp 41–56

Reid N, Carroll MC (2008) Creating trust through branding: the case of the northwest Ohio greenhouse industry. In: Stringer C, LeHeron R (eds) Agri-food commodity chains and globalising networks. Ashgate, Aldershot, pp 175–187

Reid N, Smith BW (2012) Assessing the success of an industrial cluster. Int J Appl Geospatial Res 3(3):21–36

Reid N, Gatrell JD, Ross PS (2012) Towards a relational geography of local food systems: or wicked food problems without quick spatial fixes. In: Reid N, Gatrell JD, Ross PS (eds) Local food systems in old industrial regions: concepts, spatial context, and local practices. Ashgate, Aldershot, pp 251–260

Reid N, McLaughlin RB, Moore MS (2014) From yellow fizz to big biz: American craft beer comes of age. FOCUS Geogr 57(3):114–125

Rittel HWJ, Webber MM (1973) Dilemmas in a general theory of planning. Policy Sci 4(2):155–169

Schiff R (2008) The role of food policy councils in developing sustainable food systems. J Hunger Environ Nutr 3(2/3):206–228

Shortridge JR (1996) Keeping tabs on Kansas: reflections of regionally based field study. J Cult Geogr 16(1):5–16

Thompson E Jr, Harper AM, Kraus S (2008) Think globally – eat locally: San Francisco foodshed assessment. American Farmland Trust, Washington, DC

Chapter 8
Poverty, Sustainability, & Metal Recycling: Geovisualizing the Case of Scrapping as a Sustainable Urban Industry in Detroit

Michael L. Chohaney, Charles D. Yeager, Jay D. Gatrell, and David J. Nemeth

Abstract Metal recycling is a sustainable urban industry supplied by scrap metal formally collected from primary metal producers, manufacturers, and informal collectors of discarded recyclable materials. Increased worldwide demand for metal and the limited primary metal production capabilities of developing nations, particularly China, has raised the price of scrap metal and elevated scrap metal as the United States' leading sustainable export, annually generating billions of dollars in revenue. This study theoretically situates formal scrap suppliers and scrap wholesalers (i.e. scrap yards) as elements of industrial symbiosis (Chertow MR, Annu Rev Energy Environ 25:313–337, 2000) and informal scrappers as urban mining agents (Brunner J Ind Ecol 15:339-341, 2011) whose behaviors strongly resemble that of social foragers (Giraldeau LA, Caraco T, Social foraging theory. Princeton University Press, Princeton, 2000). Using Detroit as a case study, empirical analyses, informed by field-based observations and interviews, were conducted to support each theoretical framework. Specific analysis topics included the impact of metal prices on green job creation in the metal recycling industry, the locational determinants of scrap yards, the possible role of unemployment on the informal scrapper population, the destructive impact of criminal informal metal collectors, and a

M.L. Chohaney (✉)
Spatially-Integrated Social Science, The University of Toledo, Toledo, OH, USA
e-mail: michael.chohaney@rockets.utoledo.edu

C.D. Yeager
Department of Social and Behavioral Sciences, Snow College, Ephraim, UT, USA
e-mail: charles.yeager@snow.edu

J.D. Gatrell
School of Environmental Studies, Bellarmine University, Louisville, KY, USA
e-mail: jgatrell@bellarmine.edu

D.J. Nemeth
Department of Geography, The University of Toledo, Toledo, OH, USA
e-mail: david.nemeth@utoledo.edu

J.D. Gatrell et al. (eds.), *Urban Sustainability: Policy and Praxis*,
Geotechnologies and the Environment 14, DOI 10.1007/978-3-319-26218-5_8

geo-visualization of optimal locations to informally extract scrap metal in Detroit. The successful analyses demonstrated in this chapter support the applied theoretical frameworks as a valuable platform to study the metal recycling industry and also produced new empirical and ethnographic insights on scrap metal supply chains, which deserve more intensive research as a vehicle for urban sustainability.

Keywords Urban mining • Sustainable development • Informal economy • Shrinking cities • Urban metabolism

8.1 Introduction

> For this steel we're getting eleven cents a pound, but if you take in copper you get like two dollar fifty cents a pound. It goes somewhere and gets melted. Detropia

Metal recycling is a sustainable urban industry supplied by scrap metal formally collected from primary metal producers and manufacturers, and informal collectors of discarded recyclable materials. While reusable metallurgic waste from primary production is typically recycled in-house, scrap obtained from manufacturers and informal collectors converge at scrap yards that sell sorted and processed scrap to secondary metal smelters. Detroit is a particularly intriguing case study because it contains a textured and extensive network of formal and informal scrap metal recycling suppliers; the city's remaining steel and automotive production facilities and the sizable population of informal scrappers, including a considerable proportion who illegally extract metal from the city's multitude of vacant properties. This chapter serves as a prolegomenon to productive and promising research avenues involving the geo-visualization and analysis of the socio-spatial dynamics of Detroit's metal recycling supply chain, with the purpose to broadly cover the scrap metal recycling industry as it relates to urban sustainability and stimulate future cross-disciplinary research.

Scrap metal is one of the United States' consistently growing non-agricultural export commodities estimated to produce a total of $90.6 billion in output in 2011—on par with fishing and forestry and linked with over 100,000 jobs (ISRI 2013). Geographic research on the United States' metal recycling industry's broad, textured supply chain necessitates a prolegomenon, ranging from the strategic supply and demand markets, motivation in the context of industrialization, urbanization, and sustainable local material loops. Most of the United States' scrap metal originates from manufacturers selling residual metal waste to maximize earnings and minimize materials waste, suggesting the principles of industrial symbiosis—sustainable material exchanges among symbiotic firms—may apply to the formal metal recycling industry.

In cities across the country and around the world, rapidly changing economies and industrial bases, combined with persistent poverty, have inspired many to rely on nontraditional or informal means of survival. Urban mining, the process of mining or "scrapping" waste repositories for recyclable or valuable materials, has long been a staple of the informal economies of many lesser-developed countries (Brunner 2011). However, the devastation of the Great Recession and its slow recovery has led to the rapid expansion of urban mining activities in cities around the United States (Koba 2013). Although urban mining has always occurred near the margins of American society, both formal and informal entrepreneurs are now employing aggressive and competitive strategies to collect and sell valuable parts of the urban landscape. In so doing, the metal recycling industry fills a new niche in the urban economy.

This research uses both ethnographical and empirical approaches to provide a comprehensive and robust perspective on urban mining and the metal recycling supply chain in Detroit, Michigan. The ethnographic component involving the informal side of urban sustainability consisted of a convenience sample of 15 scrappers; 5 were interviewed at length (approximately 1-h) in a semi-structured manner. The insights obtained from the scrappers were used to inform and field-test assertions tested using secondary data sources. Data sources were collected from a variety of sources, such as the U.S. Census and Bureau of Labor Statistics, and will be disclosed before each analysis.

8.2 Why Is Scrap Metal in Demand?

Scrap metal demand is derived from final demand for metal products; therefore, scrap demand (and price) increases with the use of steel, iron, copper, lead, or any of a long list of metals used for industrial or commercial purposes. The supply of prompt scrap relies on the final demand for metals as a result of industrial activity; however, obsolete scrap collection is motivated by rising scrap metal prices (Söderholm and Ejdemo 2008). The growth in international economic activity, particularly Brazil's growing manufacturing sector and China's massive construction projects, has dramatically raised metal prices. Concurrently, many mining companies are extracting lesser-quality ores from more difficult sites because most high-quality, easily mined locations have been exhausted (Ravilious 2013). Also, primary metal production has simply has not been able to keep up with consumer demand, creating a market for secondary metal produced from recyclables. Copper production, for example, has lagged behind demand to the extent that scrap prices have been as high as 90 % of the price of virgin material (Sidebottom et al. 2011). Given the difficulty, expense, and environmental impact of removing new ores from the Earth, recycling and reusing discarded metals is an understandably desirable and sustainable alternative.

8.3 Who Are Scrappers, and What Do They Do?

Informal metal collection, commonly referred to as *scrapping* and theorized as urban mining, is a significant supply source for the metal recycling industry that also produces a satisfactory and reliable income for many informal collectors, known as *scrappers*. Brunner (2011) appropriately conceptualized scrappers as urban miners because they use specialized knowledge about product composition to extract discarded materials in predominantly urban areas. Based on Chohaney's field experience observing and conversing with numerous scrappers around Detroit metropolitan area between July and October 2014, scrapping is generally conducted legally (*mainstream*) or illegally (*criminal*); however, many scrappers economically apply both approaches. Mainstream scrapping entails sifting through residential, contractor, or manufacturing refuse for valuable materials, or being commissioned to gather and dispose of waste materials from construction or demolition sites, homes, farms, among other places. Criminal scrapping involves the extraction, theft and recycling of metal from (primarily vacant) properties, particularly the coveted high-value copper and brass plumbing, wiring, hot water tanks, air conditioners, and aluminum siding are. Both scrapping approaches occur at differing geographic scales, criminal scrapping primarily occurs within blighted cities, whereas mainstream scrapping is conducted across a metropolitan area.

There is a dearth of research on informal scrapping in the United States and no known rigorous and systematic case studies. Research exploring how geovisualization might optimize urban recycling is also lacking. Most research on scrap metal recycling pertains to the trade of prompt scrap metal and consumer product lifecycles (Gruver and Giarratani 2005; Lyons et al. 2009). Masocha (2006) presented an ethnographic study on dumpsite scrappers in Zimbabwe, concluding scrapping is economically beneficial for urban poor with few employment alternatives. Gregson et al. (2012) conducted a robust, field-based study on ship-breakers in Bangladesh where she and her team identified how obsolete vessels are obtained, scrapped, and recycled and reused in a supply chain of diverse industries reliant on the material obtained from dismantling large, obsolete ocean vessels. Methodological similarities were noticed between Eynard et al.'s (2014) discussion of geovisualization as a tool to assist urban planners with sustainable development in Torino, Italy and this study's use of geovisualization to locate potentially lucrative scrapping areas in Detroit, which could be used to manage informal scrapping in the city. A section dedicated to explaining the complexities of informal scrapping in Detroit is found later in the chapter, along with a section regarding the aforementioned geovisualization.

8.4 Metal Recycling and Urban Sustainability

The seventh of the United Nation's Millennium Development Goals emphasizes environmental sustainability, including the development of efficient plans to manage municipal solid waste (United Nations 2005). Meeting such goals were touted to create jobs, relieve poverty, prevent disease, improve drinking water sources, and lessen the environmental hazards associated with the unregulated collection and transport of waste materials (United Nations 2005). In short, realized urban mining is thought to produce environmental and economic benefits.

Technological advancements are also streamlining the collection of discarded metals. Optimization using Geographic information systems (GIS) occupies the core of these advancements; in fact, many waste management companies use geospatial technologies to optimize waste and recyclable material collection (Rada et al. 2013). GIS is commonly used for sanitary landfills and waste incineration facility site selection, and has been used to model efficient recyclable collection routes (Alvarez et al. 2008). Advancements in metal processing promises to facilitate the scope of urban sustainability; for example, extracting platinum from roadside grasses, collecting rare earth elements from discarded optical cable, or using bacteria as a means of collecting nickel, zinc, and other valuable metals found in discarded urban waste (Ravilious 2013).

Scrap metal recycling is by its very nature a sustainable, environmentally-beneficial industry because it mitigates demand for newly- mined mineral ores required for primary metal production through recycling and/or reusing, as well as reducing the amount of consumer waste in material "sinks" such as landfills and incinerators; however, widespread implementation of recycling continues to be primarily motivated by economic incentives (Brunner 2011, 340). Primary metal production requires mined metallurgic ore, a scarce good involving high overhead costs; consequently, scrap metal separation and smelting have been optimized by technological improvements increasing the efficiency and quality of secondary metal production while lowering its cost to the extent secondary metal is considered a substitute for metals produced from primary ores. Due to its importance in mitigating demand for primary metal, recyclable scrap has become a globally traded commodity and a significant export for the United States, particularly to developing countries with limited ore supplies (Janssen et al. 2001).

Material reuse is the most obvious environmental benefit of scrap metal recycling. Every ton of recycled steel saves roughly 2500 lb of iron ore and 1000 lb of coal needed for virgin production. Scrap copper refinement requires 85 % of the energy as needed to produce the metal from ore (Horne 2011, 9). Production byproducts, particularly slag produced during smelting, are recycled as fill or soil conditioning in road-building. Compared to primary production, scrap metal recycling reduces CO^2 emission by 58 % for ferrous metal, 65 %, for copper, and 95 % for aluminum (Horne 2011, pp. 8–9). Recycling aluminum is particularly energy-efficient; in fact, the energy required to produce one aluminum can from virgin ore is equivalent to the energy needed to fabricate 20 cans using recycled aluminum. Additionally, ferrous and non-ferrous metals do not physically or chemical degrade over time; therefore, the benefits of recycling metals can be enjoyed indefinitely. Indeed, the lesser environmental externalities of reusing and recycling discarded materials is preferable to standard ore mining practices, and methods used to recycle metals are less energy-intensive than mining and refining ore and of comparable quality (Norgate and Haque 2010).

8.5 Growth and Change in Scrap Metal Markets

The United States is the largest scrap metal producer in the world, recycling over 135 million metric tons of materials in 2011 (ISRI 2013). Domestic demand markets exist; however, Asia—particularly China—is the world's emerging scrap metal market. Domestic scrap metal originates from two broad sources: prompt scrap, consisting of metal refuse from manufacturing or primary metal production (the latter is usually recycled in-house), and obsolete scrap recovered from discarded items and then organized and processed according to melting specifications.

Steel and iron, characterized by their significant iron content, are classified as ferrous scrap. They are also the world's most recycled metals; an estimated 72 million metric tons of ferrous scrap was traded in 2012 (ISRI n.d.). Steel is the most widely utilized industrial metal and is crucial for manufacturing; iron alloys such as wrought and pig iron are needed for construction materials and magnet production (Congressional Research Service 2008).

Non-ferrous metals do not contain substantial amounts of iron, and are used for their appreciable properties; for example, thermal and electrical conductivity (copper), low weight (aluminum), and ductility (titanium) (Fahlman 2008). Precious metals, such as gold, silver, and platinum, are also considered non-ferrous metals. These metals are conspicuously used decoratively as jewelry and for monetary purposes, but are also essential components for manufactured products. For example, platinum is used for catalytic converters in motorized vehicles, and gold's electrical conductivity renders it ideal for producing relays, connectors, and switches in electronic devices. It should be noted there are myriad grades of ferrous and non-ferrous scrap metals. Up-to-date specifications and information on non-metallic recycled materials are freely provided by the Institute of Scrap Recycling Industries (http://www.isri.org/).

Steel "mini-mills" use electric arc furnaces (EAF) to incorporate high proportions of ferrous scrap into raw steel production. In 2006, 93 mini-mills generated 57 % of total steel production in the U.S. (EPA 2009, 1). By contrast, only about 25 traditional integrated steel mills processing mined ores exist in the United States, and produce less than half the country's total steel output (Söderholm and Ejdemo 2008). Mini-mills have dominated the domestic market due to the uneconomical cost of operating large-scale integrated steel mills in a competitive global market when compared to mini-mills' lower operating costs and initial capital investment, more efficient production scale, higher productivity, and minimized labor costs gained from hiring mostly non-unionized workers (Ahlbrandt et al. 1996; Stuart 2013; ILO 1997).

The United States' copper smelting capabilities have diminished during the last 30 years due to stricter environmental regulations and deteriorating competitiveness with international producers (Jolly 2013). In 2010, the U.S. secondary copper industry consisted of 710 high-efficiency flash smelters and 123 secondary refineries, primarily producing copper for alloying; furthermore, the U.S. does not contain any secondary or reverberatory smelting facilities—down from 315 to 1526 in 1982

(Jolly 2013, 23). Nevertheless, copper scrap is still widely collect in the U.S., but generally shipped to overseas processors.

In 2012, the United States' metal recycling industry exported nearly $34 billion in commodity grade scrap products to more than 155 countries (ISRI 2013). According to 2013 U.N. Commodity Trade Statistics, the United States was the leading exporter of scrap copper and ferrous metals, with over 179 million metric tons of scrap copper (20 % of the world's scrap copper output) and 18.5 million metric tons of ferrous scrap metal sold abroad (Jolly 2013). Burgeoning metal-producing industries in China and Turkey demand the largest imports of scrap copper and steel, respectively. China imported 68 % of the world copper scrap and is the largest consumer of U.S. copper scrap, purchasing approximately 1.18 million metric tons in 2013 (Jolly 2013). Turkey imported 56 million metric tons of ferrous scrap metal in 2013, approximately 21 % of the world's steel scrap exports. Comparatively, China, the second-largest import market for ferrous scrap metal, bought about 1.18 million metric tons from the United States (Cembekli and Bayraktar 2014).

8.6 Industrial Symbiosis

Industrial ecology is the study of material and energy flows through industrial systems. However, because industrial scrap suppliers and scrap yards operate as firms, industrial *symbiosis* (IS), the study of an industrial ecological system at the inter-firm scale, most appropriately applies (Chertow 2000). Industrial symbiosis is essentially an urbanization agglomeration economy where competitive advantages due to economies of scale from waste material exchanges among proximate, functionally dissimilar firms, and made efficient by proximity and transportation infrastructure (Desrochers 2002; Marshall 1986 [1920]).

Industrial symbiosis depends on spatial interaction via relative location because transport costs between scrap metal supply sources (scrap yards) and scrap metal consumers must be low enough to economically justify the transport costs. For example, on a micro-scale, discarded aluminum rain gutters are "junk" if located an uneconomical distance from recycling facilities; however, if the facilities are nearer or more gutters are collected, the materials become a lucrative tradable to scrap yards (Desrochers and Lam 2007). On the firm-level scale, IS requires scrap yards to be within an economic cost-distance to larger wholesalers and metal producers, according to proximate relative location or accessibility to transportation infrastructure. Admittedly, scrap metal recycling does not follow the ideal definition of industrial symbiosis, which actually excludes recycling facilities; however, the industry operates by the same motivations, mainly efficient and economical resource use (Chertow 2000).

Industrial symbiosis requires at least a three to two materials exchange, meaning at least three economic entities must trade no less than two materials at several geographic scales (Chertow 2007). Metal recycling involves two material flows, ferrous and non-ferrous scrap metal, which stream one-way from scrap metal suppliers to scrap wholesalers to secondary metal production facilities operating at local, regional, national, and multinational scales. If scrap yards and industrial firms co-locate, industrial firms' negative externalities (metal waste) becomes profitable because reduced transport costs allow discarded scrap materials to be profitably sold to nearby scrap yards. This symbiosis is doubly beneficial for industrial firms because scrap metal supplies regulate the short- run metal prices; if scrap supplies increase, short-run prices for steel and copper will fall, reducing industrial firms' supply costs (Xiarchos 2005; Xiarchos and Fletcher 2009).

Detroit's informal scrappers, construed as entrepreneurial scrap metal- collection micro-enterprises, also benefit from urbanization agglomeration economies for sharing inputs and knowledge among scrappers to increases the efficiency of extraction. Informal scrappers fulfill industrial symbiosis' heuristic three to two materials exchange by trading several materials and goods, including scrap metal, antique home hardware, and valuable consumer products to scrap yards, antique dealers, and flea market peddlers located (permanently or itinerantly) near local sources in Detroit city.

The industrial symbiosis literature idealizes prescribed eco-industrial parks as goal for urban sustainability, particularly the industrial district in Kalundborg, Denmark, centered on energy sharing among a coal-fired power plant, plasterboard manufacturer, an oil refinery, and a pharmaceutical producer (Ehrenfeld and Gertler 1997). However, simple voluntary symbiotic material flows, such as those between industrial firms and scrap yards, are most common and realistically expounded upon (once identified) through non-invasive policy initiatives; in fact, planned efforts for eco-industrial parks rarely succeed (Chertow 2007, 26). In short, the principles of industrial symbiosis tend to occur when the situation is economically beneficial and has a propensity to fail when imposed by planning and policy statutes (Chertow 2007; Gregson et al. 2012, 41).

8.7 Urban Metabolism

Urban metabolism is the process of removing and disposing of "all the materials and commodities needed to sustain the city's inhabitants at home, work, and at play" (Wolman 1965, 156). Such products include water, energy sources, food stuffs, construction materials, and consumer goods. Urban metabolism conceptualizes urban areas as ecosystems where energy and materials are discarded through "metabolic" mechanisms minimizing waste products, such as recycling and reuse strategies facilitated by public policy, urban planning, and the built environment (Castán Broto et al. 2012). While urban metabolism primarily occurs at the municipal scale by informal scrap collectors, its objective mirrors that of industrial ecology—diminish the need for mining, primary production, and material sinks such as landfills and incinerators through recovery and reuse, reduce.

Industrial ecologist Dr. Paul H. Brunner's (2011) *urban mining* concept modified and optimized the idea of urban metabolism, calling for an adjustment in urban design, technology and facilities to streamline the collection and reuse of obsolete anthropogenic resource stocks within urban areas. In order for urban mining to be realized across urban areas, he advocated that material extraction should be incorporated into building designs and an increased knowledge base among the public regarding recycle materials is needed in order to obtain the critical mass necessary for optimal urban metabolism; additionally, he called for greater accessibility to facilities capable of processing a variety of recyclables (e.g. metal, plastic, and paper). Brunner states informal scrap collectors actually utilize urban mining strategies, specifically specialized knowledge about product composition and extraction skills that maximize the amount of metals recovered from obsolete anthropogenic resource stocks (Brunner 2011, 340).

Urban metabolism largely remains the domain of dedicated scrap collectors because informal recycling operates at an unwieldy economy of scale for the general public, and disassembling the urban built environment is usually dirty, dangerous, and time-consuming. Consequently, most homeowners leave recyclable objects at the side of the road to be picked up by informal scrap collectors or "junk-men." The first obstacle for residential metal recycling is an unwillingness to accumulate enough scrap metal needed economically justify a trip to the scrap yard, likely because large quantities of scrap metal quickly become voluminous and unsightly. Greater numbers of accessible scrap yard locations or routine municipal scrap collection may be the solution; however, the informal economy has filled this service gap by advertising for junk or scrap pick-up services at little or no cost. The formal scrap metal industry is also beginning to adopt this strategy; one Detroit scrap yard offers pick-up services and finds it a competitive advantage (M. Chohaney, interview, July 15, 2014).

Additional barriers include a lack of awareness and education on how to discern different types of metals and their grades and the unwillingness to undergo the dirty, dangerous, and demanding tasks involved in dismantling, extracting and separating postconsumer materials. Individuals can be educated on product deconstruction by online videos outlining metal collection strategies and dismantling how-to's and tips publicly shared on Youtube by scrappers. However, the intensive labor requirements remains a roadblock; therefore, until these obstacles are addressed with labor-saving technologies or sufficient numbers of processing facilities, urban mining will continue to be a dirty, labor-intensive process only few are willing to undertake.

Scrap metal collecting and recycling is the most pervasive and profitable form of urban mining/metabolism. Obsolete or non-functioning ("junk") consumer goods are frequently collected and recycled, varying from appliances and automobiles to smaller metal items like bed frames, bicycles, and leftover household metal products such as piping or wiring. Construction sites often recycle aluminum window frames, vents, siding, and rain gutters, as well as sheet metal. Interestingly, obsolete farm equipment is also sought after by informal scrap collectors for the materials' size (and thus value), perhaps jeopardizing the relevance of "urban" in urban metabolism. Obsolete or "junk" buildings are a major—though largely illegal—source of scrap metal for informal collectors in Detroit, and have been identified as reservoir for scrap iron and copper since at least the 1940s (Ostrolenk 1941, cited in Desrochers and Leppälä 2010, 37).

8.8 Scrap Yards: Essential Nodes for Urban Metabolism

The United States contains more than 3500 scrap processors that supply approximately 100 domestic steel mills and 600–700 domestic foundries (Aylen and Albertson 2006, 203–4). Scrap processors are commonly known as scrap yards, which operate within a hierarchical network of wholesalers that purchase, sort, and process scrap metal into usable material for manufacturing or metal production (Fig. 8.1). Most scrap metal is purchased or retrieved from industrial firms in contractual agreement with a scrap yard; however, scrap yards also buy discarded materials from construction/demolition contractors and informal scrappers. Scrap yards' final products are dependent on the yard's processing capability, which increases in complexity according to the size of the yard due production efficiencies arising from economies of scale. Smaller yards typically sell sorted and cleaned (i.e. non-metal material removed) scrap metal, though some have shredding or compacting equipment. Large yards produce briquettes, bundles, or bales of scrap sold to steel mills, and may shred large recycled items (e.g. automobiles) for advanced sorting using electro-magnets, among other methods (Fig. 8.2). Final customers for sorted and processed scrap are multinational wholesalers, smelters, refiners, ingot makers, and foundries.

Fig. 8.1 Grapple cranes loading ferrous scrap metal into a compactor/baler, River Rouge, Michigan

Fig. 8.2 Bales of scrap metal at a large capacity scrap yard, River Rouge, Michigan

Scrap yards sort ferrous or non-ferrous metals according to its commodity grade. Sorting is simple for industrial scrap because the composition of is certain; however, obsolete scrap must be manually differentiated, cleaned, and re-sorted—samples are taken from each bundle before shipment to ensure a consistent composition. More precise non-ferrous metal sorting is sometimes conducted at large-scale yards along conveyor belts using magnetic Eddy fields to discern metal composition (Stessel 1996). Another popular automated sorting method uses high-pressure air flows to remove non-metallic materials from a magnetized surface. Sorted and cleaned scrap is compacted to commodity-grade specifications, purchased according to weight-price, and shipped via truck, rail, or container ship to secondary metal mills and smelted into new metal products.

Brunner (2011) postulated scrap processing centers most efficiently locate near source materials to reduce transportation costs (340). Smaller scrap wholesalers cluster near higher-capacity yards, with whom they presumably trade and share common inputs, as shown in Fig. 8.3, which displays the distribution of Detroit's scrap yard's proportionally according to square footage (measured using Google Earth). It appears smaller scrap yards are more dispersed than larger yards, which cluster in industrial areas, likely the result of shared proximity to rail transportation needed to access to primary scrap metal markets.

Scrap yards typically use tractor-trailers or open-top "gondola" railcars to transport scrap metal; multinational wholesalers use "handysize" dry bulk vessels or container shipping to transport hundreds of thousands of tons of scrap metal to foreign markets (Parker 2000). Tractor- trailers are used to transport sorted scrap

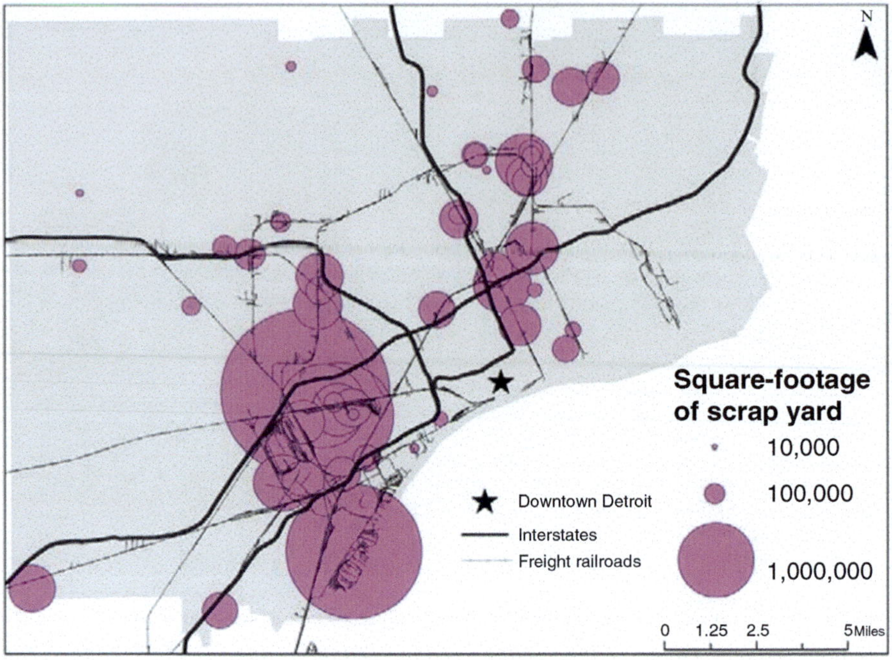

Fig. 8.3 Proportional square footages of scrap yards in Detroit. Michigan

among proximate yards, typically from smaller to larger yards; however, large wholesalers also use tractor-trailers to transport scrap bales to nearby customers. Gondolas are needed for economical regional shipping, but their prohibitive price of railroad spurs limit the possibility of rail shipping for most yards. Approximately 1.7 million tons of scrap material (7.8 % of all shipped tonnage) originated on Michigan's rail system in 2012; unfortunately, however, origin-destination data for scrap metal was not available (Association of American Railroads 2014). Commodity-based trucking data was also unobtainable.

While freight and trucking data was unavailable, aggregated 2010 U.S. Army Corps of Engineers Waterborne Commerce waterway shipping data for ferrous and non-ferrous scrap metal indicated Detroit did not import or export any scrap metal via waterway. Outgoing ferrous scrap shipments originated from Michigan ports near Holland and Benton Harbor on the West side of the state. The destination was Sault Saint- Marie, Ontario, likely a stop on the way to a larger East Coast port. It is possible ferrous scrap from Detroit and Southeast Michigan is transported by rail to Chicago and Cincinnati, the largest nearby ports where high tonnages of ferrous scrap was shipped from, either to the East Coast along the Ohio River or south to the Port of Orleans.

Non-ferrous scrap was also not imported or exported via waterway from any Michigan port in 2010. The nearest port shipping non-ferrous scrap was Cincinnati, Ohio, transporting to metal Gary, Indiana, where it was likely used for making steel alloys or shipped via rail to West or East Coast ports. However, scrap may have been shipped directly to Gary via railway, for waste and scrap materials comprised the majority of Michigan's rail-based exports to (presumably Gary) Indiana; similarly, these non-ferrous metal comprised 28 % of freight rail shipments to Pennsylvania (HNTB Corp. 2011, 37). While railway and trucking freight data are needed to fully explore the aforementioned speculations regarding Detroit scrap export network, the majority (if not all) scrap from Detroit and southeast Michigan can be safely presumed to be shipped locally by truck or regionally via railway.

8.9 Scrap Yard Site Selection

Scrap yards theoretically locate where they can take advantage of efficiencies gained from industrial symbiosis, meaning they locate near scrap suppliers and areas easily accessible to transportation networks. Nearness to industrial firms ensures consistent access to nearby buyer markets and substantial supplies of profitable, low-residual scrap facilitated by familiarity and the low transportation costs (Chertow 2007; Gruver and Giarratani 2005, 315). Accessibility to interstates optimizes trucking to a greater number of local customers; immediate railway access may also motivate the location of scrap yards, particularly those operating at a large scale.

A reverse site selection analysis was conducted to measure the relative significance of scrap yards' theoretical site selection criterion. First, a lattice comprised of 586 cells with individual areas of 0.703^2 miles (the median area of the study area's Census Tracts) was placed over the study area. Census Tracts were integrated into the lattice by adjusting the data proportionally to the portion located in the grid cell. There were a total of 55 scrap yards (including auto-recycling processors) located in 7 % of the grid cells, resulting in a substantial zero count (93 %) and over-dispersion—violating the assumption that the conditional variance of the dependent variable is equal to the conditional mean. Therefore, to address these issues, the zero-inflated Poisson model was chosen, which includes two components measuring the two evident data-generating processes—one according to the locational determinants and another the result of an unchanged exogenous process explaining the zero counts. Industrial zoning regulations would have been ideal for this analysis' zero-inflated component; unfortunately, zoning shapefiles were unavailable for most of Detroit city's surrounding municipalities. Instead, the total number of housing units (from the 2010 U.S. Census) was used to explain the zero-inflated component because residential areas are typically located apart from industrial zones. The four best-fitting models were reported to assess the relative magnitude of different combinations of theoretically relevant location determinants.

The relevance of industrial symbiosis was measured by relating the scrap yard counts to the number of industrial firms in the same grid cell, within a half or one-mile of the grid cell centroid, or the distance to the nearest supplier. Typical prompt scrap suppliers, such as manufacturing firms, metal works, and steel-mills were considered as scrap suppliers. Demolition contractors were measured as a separate category of scrap supplier. Industrial locations were gathered via GoogleEarth using the search results for the keywords "industrial Detroit," "manufacturing Detroit," and "steel Detroit;" duplicates and irrelevant results (e.g. brewers and paper recyclers) were removed. Demolition contractor locations were gathered from the keyword "demolition Detroit" and were also sifted for relevancy.

Transportation infrastructure measures included rail yards, intermodal terminals and interstate ramps. Interstate ramps were extracted from the U.S. Census' 2013 Tigerfile database. Accessibility was measured as the grid centroid distance to the nearest transportation network; alternatively, grid cells were dummy-coded dependent on whether they fall within or beyond 1 or 3-mile buffers of each ramp or rail facility. Intermodal facilities were geocoded based on addresses provided by the Michigan State Rail Plan (HNTB 2011). Rail yards were extracted from the Chicago Transportation Authority's North American rail network (http://www-cta.ornl.gov/transnet/RailRoads.html).

The models' results can be found in Table 8.1. The statistically best-performing model (Model 1) explained approximately 25 % of the variation in scrap yard

Table 8.1 Zero-inflated Poisson regression model results

Count of scrap yards	Model 1 (robust std. errors)	Model 2 (robust std. errors)	Model 3 (robust std. errors)	Model 4 (robust std. errors)
Proximity to demolition contractor (miles)	−0.4238841 (0.159217)**	−0.3386988 (0.1400781)**	−0.4086831 (0.1448114)**	−0.3500183 (0.1611283)**
Proximity to an intermodal terminal (miles)	−0.3661527 (0.1048141)***		1.685841 (0.3692479)***	
Proximity to a railroad yard (miles)				−0.4786372 (0.1380503)**
Within 1 mile of an intermodal terminal		1.610471 (0.3652683)***		
Within 1 mile of an interstate ramp	0.5582899 (0.3718899)	0.6108776 (0.3725339)		0.5058443 0.3704485
Proximity to interstate ramp (miles)			−0.0800605 (0.3109868)	
Number of industrial firms within a half mile	0.4383505 (0.0492699)***	0.4030004 (0.0402024)***	0.3842172 (0.0348265)***	0.2400568 (0.0581153)***
Hundreds of households (zero-inflated component)	0.2059824 (0.0638465)**	0.2039476 (0.0674543)**	0.2035279 (0.0684078)**	0.2340442 (0.0855441)**
Nagelkerke R^2	0.251	0.25	0.244	0.241
AIC	0.521	0.522	0.525	0.527

p>0.10,**p>0.05,***p>0.000

counts, and produced the lowest Akaike Information Criterion (AIC). The zero-inflated portion, the number of housing units (per hundred), was also statistically significant with a positive coefficient in all the models. In Model 1, an additional 100 homes increases the probability of a zero-count by factor of 1.22, supporting residence counts as a meaningful substitute for zoning regimes.

Model 1's results indicate industrial symbiosis is an important factor for scrap yard site selection. However, proximity to intermodal terminals was the only statistically significant transportation variable; the number of scrap yards decreases by a factor of 0.65 for every mile closer a grid cell centroid is to an intermodal terminal. Nearness to interstate ramps was not statistically significant, likely because it is uneconomical to use trucks to ship heavy, relatively low-value scrap long distances via interstates. Tractor-trailers are mostly used for locally transporting relatively small amounts of sorted loose scrap from smaller to larger yards, and by larger yards to ship bundled or shredded scrap metal to smelters (M. Chohaney, interview, September 24, 2014; Lyons et al. 2009). This suggests Detroit's scrap yards export most of their materials regionally via rail; however, intra-city scrap shipments should be considered for future analysis to measure the importance of potential material loops or urban sustainability among automobile manufacturers, scrap yards, and steel mills that refine scrap metal into finished metal products that they sell back to city's automobile manufacturers.

How scrap is shipped by rail is unclear—only one of the four best-performing models produced statistically significant coefficients for the rail transportation variables—Models 2 and 3 found intermodal terminals positive and statistically significant, insinuating Detroit's scrap yards may use intermodal containers instead of the traditional gondola freight cars made for dry bulk goods. However, this contradicts field interviews and observations confirming Detroit scrap yards with rail access load scrap onto gondolas, suggesting the significance of intermodal terminals may be attributed to these facilities' tendency to be located in heavily industrialized areas where many businesses own railroad spurs. Therefore, intermodal terminals may be acting as a proxy for rail spur ownership, an essential element for rail transport that public rail data does not accurately report. Still, while the use of intermodal containers for shipping scrap metal is unrealistic in Detroit, smaller scrap yards along the West Coast have adopted container shipping directly to buyers as a cost-effective shipping option, favoring independent brokering and shipping to buyers over selling scrap to nearby highly-accessible and high-capacity wholesalers at diminished rates (Iritani 2005; Marley 2007).

Industrial symbiosis most significantly influenced scrap yard site selection. One additional industrial firm within and near (0.5 miles) each cell's centroid increased the number of scrap yards by a factor of 1.55. This result affirms the significance of industrial symbiosis because symbiotic firms, in this case scrap yards and industrial producers, exhibited a tendency to co-locate—presumably facilitating materials exchanges (Shi et al. 2010). Nearness to demolition contractors produced a negative coefficient, suggesting demolition firms are not symbiotic with scrap yards in the same manner as industrial firms. Social proximity between demolition firms and scrap yards may be more relevant because material collection from demolition contractors occurs at dispersed work sites; therefore, proximity to a demolition company's front office does not create any discernible efficiency (Gregson et al. 2012, 41).

In addition to more accurate data on rail spur locations and ownership and intra-city scrap metal transportation, a Poisson regression model accounting for spatial dependence should be used in future analyses to correct or spatial dependence that violates the model's assumption of independent observations. The Moran's I of the dependent variable was 0.13 and all of the explanatory variables featured substantial and statistically significant levels of spatial autocorrelation; however, the Pearson residuals' Moran's I was not statistically significant, and the raw residuals only resulted in a modest, statistically significant Moran's I of 0.09.

8.10 The Negative Externalities of Scrap Metal Recycling

The scrap metal recycling industry is conceptually environmentally-beneficial and fundamental for sustainable urban growth; however, in practice, it is often not the idealized "green" activity often portrayed. Most scrap yards are future brownfield sites as a result of contamination from decomposing metals and chemicals and oil leaks from appliances and motorized vehicles while piled in large scrap heaps (Blake et al. 1987; Yin et al. 2007). More interestingly, however, is the significant environmental damage caused by the criminal component of Detroit's informal scrappers, whose environmental impact and working conditions strongly resemble the infernal task landscape of ship-breakers in the Global South.

The resemblances between Detroit's criminal informal scrappers and ship-breakers in the developing world supports Gregson et al.'s (2012) assertion that industrial symbiosis can manifest as messier, dirtier, and more dangerous enterprises than the idealized, Euro-centric case studies often presented in the literature (39). Ship-breaking is the process of disassembling obsolete ships and vessels for metals and materials; therefore, *city-breaking* is a term appropriate to describe the extraction of valuable materials from obsolete urban infrastructure and built environments, particularly abandoned or vacant structures in blighted areas. However, city-breaking only pertains to criminal informal scrapping because it is metal *theft* from vacant or abandoned properties.

While Detroit is located in a developed country with relatively stringent environmental regulations and standards, the city provides a convenient metaphor for describing how humans in a post-industrial era can foul their own nests. In 2014, 10,220 parcels (3 % of the Detroit city's parcels) were described as illegal dump-sites, according to the Motor City Mapping parcel survey—a private-public venture motivated by enumerating and appraising every parcel in Detroit (http://d3.d3.opendata.arcgis.com/). Abandoned properties and vacant structures also commonly resemble or serve as such dump sites—an aesthetic immortalized by the Heidelberg Project neighborhood art exhibit on the city's East Side (www.heidelberg.org).

Criminal scrappers in Detroit tend to operate under the mantra of "rip and run," basically extracting the most scrap from (mostly) vacant properties in the shortest

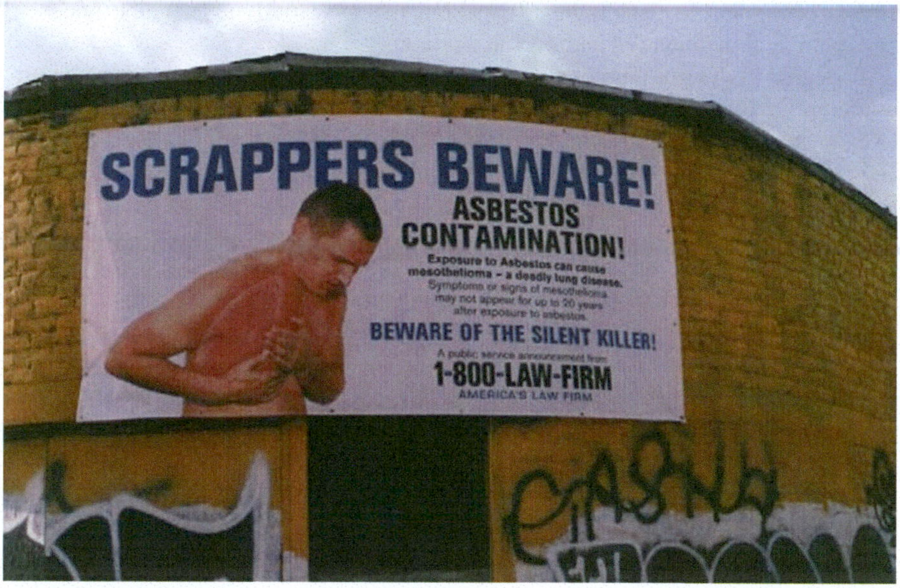

Fig. 8.4 Advertisement posted by a local law firm on an abandoned and mined property along Michigan Avenue, Detroit, Michigan (Photo by Chohaney)

duration (M. Chohaney, field notes, September 25, 2014). The "rip" phase entails rapidly dismantling structures, producing a mass of rubble and debris and jeopardizes the building's structural integrity; additionally, this stage is conducted with no concern for hazardous materials, resulting in exposure to asbestos and lead-based paint and mercury from old appliances (see Fig. 8.4). Active natural gas and water lines are frequently damaged, causing structure fires and foul-smelling, mosquito-producing pools of stagnant water. Environmental concerns are minimal, if considered; the only major consequence for creating piles of debris or unsafe structures are disapproving neighbors who may jeopardize the ability to work in the area (M. Chohaney, interview, July 18, 2014). Similarly, Bangladeshi ship-breakers often improperly drain fluids from vessels, causing oil and hazardous materials leaks into the water and onto the beach, and also regularly exposed to asbestos and other toxic materials (Gregson et al. 2012).

Low population density in Detroit's severely blighted neighborhoods has diminished residents' will to fight scrappers; additionally, widespread poverty has facilitated scrapping in general as well as bribery involving residents who agree to look the other way while scrapper(s) operate throughout their neighborhood. Nevertheless, numerous neighborhood and non-profit organizations remain committed to fighting blight through neighborhood cleanups and anti-blight efforts (e.g. boarding-up windows and doors) to prevent intrusion by scrappers and others—one such organization being the Detroit Blight Busters (http://www.mcbbdetroit.com/). Detroit's financially-strained police force cannot feasibly monitor and arrest enough scrappers to deter others, exposing the most outlying neighborhoods to environmental degradation and overall decline.

A dearth of gainful employment for low-skill workers in and immediately surrounding Detroit has motivated many individuals to risk health, safety, and jail time to earn ample, tax-free incomes as self-employed informal scrappers. Successful and safe urban mining, particularly dismantling structures to extract structural steel among other materials, requires a "degree" in what urban miners call "scrapper physics;" however, these skills are not always sufficient to avoid hazards (M. Chohaney, interview, July 11, 2014). One scrapper, "J," broke his leg after falling through the floor of an old house while removing a cast-iron bathtub from an upstairs bathroom. Similarly, another scrapper, "K," recalled a peer who became trapped underneath the collapsed walls and remnant roof of a section of the Packard Plant that weakened after recently catching fire. The man was reportedly rescued by a nearby scrapper and suffered several broken ribs; however, he did not seek immediate treatment to avoid being arrested and assessed a $5,000 fine for metal theft. Bangladeshi ship-breakers are also regularly exposed to similar hazards, such as falling debris and fatal or traumatic equipment missteps; therefore, they too must carefully learn (through experience) informal demolition skills (Gregson et al. 2012).

8.11 Scrapping as a Green Job Creator: An Empirical Analysis

Despite the apparent environmental degradation produced by metal recycling industry, the Occupational Health and Safety Administration nevertheless labels it a "green" industry (OHSA n.d.). Green occupations are defined as professions in industries that minimize or altogether avoid the generation of all forms of waste and pollution (UNEP 2008). Primary (LME) and scrap steel and copper prices influence each other bi-directionally; therefore, primary prices approximately reflect overall demand for the metals (Xiarchos 2005; Xiarchos and Fletcher 2009). Due to the surprising lack of research on the economic relationship between scrap metal demand and employment growth in the U.S.'s scrap metal recycling industry, a brief regression analysis was conducted measuring the effect of scrap metal price and growth in employment related to metal recycling in Michigan and the United States (for an economic impact study for the U.S.'s industry, see ISRI 2013; for international studies on green jobs on employment, see Borel-Saladin and Turok 2013; Cai et al. 2011).

Employment growth rates for recyclable material merchant wholesalers (NAICS 423930), which includes traditional scrap yards and automotive wrecking yards, were chosen for analysis because they represent essential scrap metal suppliers and are supposed to closely depend on metal demand markets. It is expected that employment growth rates of recyclable material merchant wholesalers (RMMW) will increase in a response to greater metal market demand, measured as market prices for steel and copper.

Ten years (2003–2012) of annual RMMW employment data were collected and analyzed to measure how employment growth rates respond to increased market demand (Table 8.1). Log-normalized employment growth rates (based on a 2002 employment base year) and London Metal Exchange (LME) copper and steel prices were used as the models' variables. Price time lags were used to account for delayed responses to market demand. Models were computed for Michigan and the United States to compare their relative response to demand. Classical linear regressions (CL) were computed to measure the marginal impact of the change in metal prices on RMMW employment growth rates. Huber-White sandwich robust standard errors were implemented to account for non-normality and heteroskedasticity typically problematic in small datasets; furthermore, the data was statistically significantly ($p=0.05$) temporally stationary.

The CL regressions produced relatively high goodness-of-fit measures. Changes in log-normalized steel prices explained approximately 45–80 % of variation in employment growth, and copper growth rates explained about 47–84 % of employment growth variability. The results indicate a 1 % increase in metal prices produces between 0.12 % and 0.25 % increases in RMMW employment growth rates, supporting the supposition that scrap yards create green jobs in response to greater metal market demand. Model results can be found in Table 8.2. The Michigan models' coefficients and goodness-of-fit were noticeably lower than that of the national model, suggesting structural differences between the state's and nation's RMMW industry structure. Furthermore, regressions using total employment growth rates for Michigan and the United States as dependent variables were also tested to ensure RMMW employment growth rates were not just the result of total employment growth. The results indicated all metal price measures were statistically significantly but negatively associated with total employment growth in Michigan, likely suggesting RMMW was a growth industry during the state's overall economic

Table 8.2 The relationship between employment growth rates and market demand for steel and copper

RMMW employment growth rates	USA-no lag (robust std. errors)	USA- lagged price (robust std. errors)	MI-no lag (robust std. errors)	MI- lagged price (robust std. errors)
(ln) Steel price	0.253	0.192	0.145	0.121
	(0.036)***	(0.036)**	(0.041)**	(0.017)**
Adjusted R^2	0.796	0.692	0.452	0.491
AIC	−3.107	−2.697	−2.762	−2.836
(ln) Copper price	0.177	0.157	0.106	0.107
	(0.026)***	(0.020)***	(0.021)**	(0.019)**
Adjusted R^2	0.742	0.836	0.466	0.72
AIC	−2.875	−3.329	−2.789	−3.435

p>0.10, **p>0.05, ***p>0.000

downturn; however, metal prices did not have any statistically significant effect on total employment growth for the United States.

Michigan's dampened market response to demand is possibly as result of the state's mature RMMW industry as a result of its historic industrial and manufacturing base, as well as a reflection of the state's weakened economy during the 2000s. Intuitively, the state's falling industrial output may have diminished prompt scrap metal supplies, hindering RMMW's ability to meet increasing demand and production potentials. Alternatively, Michigan firms may operate at high efficiency, necessitating fewer newer hires when demand increases; however, scrap yard productivity (annual sales divided by employees) rose by a factor of approximately 2.2 in both the U.S. and Michigan between 2002 and 2007, according to the U.S. Economic Census, suggesting a difference in labor force skill level and available technology did not significantly hinder or promote employment in the state. Productivity increases corroborate with information provided by managers of a large corporate scrap yard in Detroit who agreed the large profits earned during the, mid-2000s metal demand spike were used by most yards to update or purchase new equipment and not to hire additional employees (M. Chohaney, interview, October 7, 2014).

Additional employment is not necessary during increased demand periods because prompt scrap supplies are dependent on industrial output, meaning scrap yards are only able to obtain a relatively fixed amount of prompt scrap; furthermore, scrap yards have a limited range to search for new prompt scrap sources due to limiting scrap metal transportation costs and existing contractual agreements between industrial firms and scrap wholesalers. Obsolete scrap supplies do increase with higher metal prices as informal scrappers are incentivized to collect and sell scrap; however, increasing obsolete scrap supplies did not apparently have a discernible effect on employment because the average sizes of RMMW firms were essentially identical, both ranging by about 2.21 employees from 2003 to 2012 (Damuth 2011). Therefore, since RMMW are not expanding, employment gains are likely derived from newly-established RMMW in response to greater market demand. This asser-

Table 8.3 The relationship between establishment growth rates and market demand for steel and copper

RMMW establishment growth rates	USA-no lag (robust std. errors)	USA- lagged price (robust std. errors)	MI-no lag (robust std. errors)	MI- lagged price (robust std. errors)
(ln) Steel price	0.185	0.148	0.075	0.064
	(0.051)**	(0.040)**	(0.017)**	(0.014)**
Adjusted R^2	0.608	0.596	0.565	0.632
AIC	−2.852	−2.822	−4.47	−4.639
(ln) Copper price	0.123	0.115	0.056	0.052
	(0.037)**	(0.027)**	(0.015)**	(0.008)***
Adjusted R^2	0.492	0.641	0.609	0.757
AIC	−2.593	−2.938	−4.577	−5.053

p>0.10, **p>0.05, ***p>0.000

tion was tested and establishment growth rates were used as the dependent variable and regressed on log-normalized metal prices. See Table 8.3 for the results. The coefficient magnitudes diminished compared to the corresponding employment regressions, but the models for Michigan fit noticeably better, supporting the latter statement that market demand was met by founding new firms rather than via expanding existing establishments, which was a more important source of employment growth among RMMW in Michigan than RMMW in the United States.

Deterrents to growth, such as regulatory barriers and prohibitively high expansion or equipment upgrading costs might also explain Michigan's weaker employment response. Corporate purchasing, restructuring, and optimization of small, family-owned scrap yards should also be recognized as a potential hindrance to employment growth and analyzed in future research to measure governmentally-imposed limitations on the industry. In addition to a larger sample size, future studies should relate price with the output of RMMW to more precisely measure how firms respond to greater market demand. Deciphering the organization dynamics behind RMMW responses to market prices, including modifying output levels, expanding capacity, or upgrading equipment may provide more informative answers about the employment structure of RMMWs and should be considered for future research.

8.12 Mining, Scarcity, & Sustainability: Evidence from Detroit

Informal metal scrapping has been traditionally motivated by economic hardship and/or inspired by the perceived and/or rumored success of others; furthermore, recruitment by acquaintances and/or even perfect strangers is not uncommon at scrap yards. In Detroit, the situation is no different—high unemployment rates and rising metal prices have created a situation where scrapping is worth the risks to a growing group of people. While this situation has likely stimulated the growth of informal scrapping across the United States, the abundance of scrap metal housed in Detroit's multitude of abandoned properties, coupled with the region's high population density, makes Detroit an especially viable place to earn a living as a scrapper following either criminal or mainstream approaches.

Detroit city has had the highest unemployment rate among the top 50 largest U.S. cities for the last 5 years. However, the number of unemployed Detoiters is falling—the Bureau of Labor Statistics reported 62,000 people were unemployed in 2012, down from 94,000 in 2009. Regionally, unemployment in the metropolitan area was just below 200,000 in 2012 after peaking at approximately 300,000 in 2009. The continual demise of manufacturing in Detroit, particularly the plight of the "Big 3" American automakers, greatly contributed the staggering unemployment figures. In fact, according to the U.S. Census' County Business Patterns, the Detroit MSA lost

Fig. 8.5 Metal prices and unemployed persons in Detroit city, 1998–2012. The initial spike in metal prices (2003) is indicated by the *first vertical line*, and the start of the Great Recession (2008) is indicated by the *second vertical line*. The period between the lines is likely the first scrapper influx, followed by massive entry during the Great Recession

approximately 20 % of its manufacturing jobs—49,000 total—during the Great Recession (between 2006 and 2012).

As seen in Fig. 8.5, scrap metal prices have risen along with city unemployment for the past 15 years. Copper prices strongly and statistically significantly ($p=0.05$) correlate with the city's unemployment rate, producing a correlation coefficient of 0.66. Steel statistically significantly correlates with a coefficient of 0.65. The intensity of co-integration between changes in metal prices and unemployment numbers could potentially be used to estimate the intensity of scrapping or scrapper entry rates; furthermore, given a baseline scrapper population approximation, this rate could be used to estimate the total informal scrapper population. When asked to describe the expansion of scrapping in the region, one Detroit scrapper, "K," stated there are currently more scrappers than he remembers in his over 10-year scrapping experience, and the numbers have accelerated over the past 5 years (M. Chohaney, interview, July 17, 2014). Apparently, fluctuating metal prices during the Great Recession did not deter many scrappers, suggesting financial necessity outweighs the cost of the metal price instability.

The large number of underemployed and unemployed former blue-collar workers in Detroit has diversified the informal labor force. Laid-off or disabled workers may end up as mainstream or criminal scrappers as a last resort attempt to earn a living, or to supplement their federal allowance; additionally Detroit youths have also turned to scrapping to earn an income. One young former scrapper from Detroit city, "D," mined homes to earn a disposable income as a teenager with his close friends until they were arrested for mining an empty, for-sale home in neighboring Dearborn. He described scrapping as a low-risk illegal economic activity (aka hustle) in comparison to alternatives, specifically drug-trafficking (M. Chohaney, interview, August 28, 2014).

Scrapping can be a lucrative entrepreneurial business; a sizable amount of scrap can be collected in only a few hours of work. Furthermore, scrapping is an informal, all-cash operation not subject to taxes, regulations, or other bureaucratic barriers entry. The contacted scrappers reported average weekly earnings ranging from $200 to $1500 (M. Chohaney, field notes, September 25, 2014). The lowest earner was a mainstream scrapper, while the most successful was 20-year veteran criminal scrapper. All the respondents agreed they could not find equally-lucrative formal employment, considering the time spent working per week—roughly 20 h on average—allowing more time for family and side-jobs than traditional hourly employment (M. Chohaney, field notes, September 25, 2014).

Poverty, blight, and a notoriously dysfunctional and corrupt bureaucracy perpetuates criminal scrapping in Detroit insofar as it creates newly minable structures and reduces residents' ability and motivation to improve their neighborhood as well reducing any effective deterrent to scrappers' illegal and successful practices. Scrapping prohibits residential revitalization in blighted areas because many housing renewal sites are broken into and mined if left unguarded. One Detroit realtor confirmed foreclosed properties in blighted neighborhoods are stripped regularly, and even suggested that prospective buyers should factor scrapper damage into rehab costs (personal correspondence, 10 Jul 2014). One might suspect re-development plans for larger vacant properties, particularly the heavily-scrapped Packard Plant, have been negatively influenced by scrapper-induced damage and partial demolition. Once local communities capitulate and accept scrapping in their neighborhood, the immediate and surrounding areas soon become victim to intensive city-breaking (Neavling 2014).

Criminal scrapping is essentially metal theft; however, not all criminal scrappers limit themselves to vacant or foreclosed properties. In-use residential or municipal fencing, siding, telecommunications cables, and catalytic convertors are commonly stolen items; still, all metal items are targets, including beer kegs, bronze statuary, and café furniture (Kooi 2010). Scrap yards and utility companies experience direct theft as well; in fact, a successful deception scam was specifically devised for this purpose, entailing individuals posing as Native Americans and asking businesses and utility companies for scrap metal donations for "traditional craft-making" (Cruz 2014; Goodwin 1993). Such approaches, when executed without detection, are more lucrative, less labor-intensive, and can be completed in broader geographic contexts than city-breaking. Metal theft has become so rampant that law enforcement has established an online database of law enforcement-reported metal theft incidents nationwide to foster collaboration for more effectual property recovery (http://www.scraptheftalert.com/).

Metal theft has been particularly troublesome in Detroit; scrap metal theft costs the city up to $25,000 per day (Halcom 2014). Many steal valuable active city infrastructure, including manhole covers, mausoleum doors, gates, and windows, copper fire hydrant valve covers, and copper wiring from street lights, transformers, and electricity grid infrastructure (Halcom 2014; Neavling 2014; Wolfe 2011). Active electricity wires are another hazard for criminal scrappers; for instance, a scrapper was electrocuted to death after attempting to remove electrical wires inside a manhole (Jackson 2014).

Rampant metal theft and pervasive criminal scrapping has prompted Michigan to enforce added security measures imposed when selling scrap metal in the state. Such procedures include fingerprinting, photographing sellers every visit, recording customer license plate numbers, photocopying identification cards, requiring written descriptions of the supplied scrap (including its origin), written affirmation that the individual is the owner of the scrap or the is authorized to exchange the metal, and written acknowledgement the individual has not been convicted metal theft (Scrap Metal Regulatory Act 2014). The law also implemented a cash payout cap, stating trade-in values appraised more than $25 must be paid by check, money order, or an electronic payment card that can only be used on-site and must record a digital or electronic image of the cash and receipt redeemer (Scrap Metal Regulatory Act 2014). The latter policy is particularly inconvenient for criminal scrappers and scrappers without bank accounts or consistent home addresses; therefore, many commonly bypass the rule by selling small amounts of scrap to many different yards, addressing the check to an acquaintance's mailbox, or exchanging the check with an informal lender providing immediate cash; furthermore, convicted metal thieves often pay someone a nominal fee to trade their scrap (M. Chohaney, field notes, September 12, 2014). While effective, these workarounds lower scrappers' productivity and reduce scrap yards' metal intake, particularly for valuable non-ferrous metals. As a result of these policies, stolen items will probably be more frequently pulverized or melted to obfuscate its origins, an additional operating cost for criminal scrappers that might deter metal theft.

8.13 Scrapping as Social Foraging: An Exploration

The manner, motivation, and distribution of informal scrapping activities strikingly resemble social foraging behaviors outlined in the behavioral ecology literature (Giraldeau and Caraco 2000). Foraging is a tactic or strategy used to obtain the resources necessary for survival, and is theorized as either resource-optimizing or social behavior. Social foraging is centered on how groups of social animals work together or interfere in the search for necessary resources; in contrast, optimal foraging theory states individuals forage to obtain the maximum amount of resources using the least amount of energy. Informal scrapping more closely relates to social foraging theory because scrappers extract resources interdependently; that is, scrappers or scrapper crews must take the activity of their peers into consideration when choosing a resource patch (e.g. refuse piles or an abandoned property) and extraction approach. Understanding informal urban mining in a social foraging context assists in organizing their foraging group sizes, strategies, information-sharing, and spatial interactions—insights that may be used to develop efficient formal urban sustainability strategies.

Social foraging research is usually conducted using a game-theoretical or experimental approach; however, because observing all informal scrapping activity is infeasible and adopting a game-theoretical framework is beyond this study's scope, principal SFT concepts will be briefly ethnographically contextualized and informed by relevant anecdotes from scrappers and field observations. While non-human species are often the subjects of behavioral ecology research, humans have also been used as case studies (see King et al. 2011; Mann and Kiflawi 2014; Sumner and King 2011; Winterhalder and Smith 1981). These exploratory results are intended to offer a working theoretical behavioral explanation for informal scrapping activities and motivate future applications of SFT to modern human foraging behaviors.

8.14 Categories of Social Foragers

Social foragers are broadly categorized as producers, scroungers, and opportunists. Producers generate their own resources, whereas scroungers manage to seize resources from producers, and opportunists flexibly act as either depending on which opportunity is most profitable (Winterhalder and Lu 1997, 47). According to this topic, resources include scrap metal and valuable found objects, such as antique fixtures or desirable consumer goods.

Most informal and mainstream scrappers can be accurately defined as opportunists; however, mainstream scrappers were more frequently observed acting as scroungers (i.e. dumpster-divers writ large). Mainstream scrappers *produce* resources when hired or granted permission to dismantle properties; otherwise, they *scrounge* through refuse piles prepared by pedestrians (municipal garbage cans), residents (personal trash bins/apartment dumpsters), or contractors (site dumpsters). Criminal scrappers produce their own scrap by locating, opening, and mining resource matched, such as seal vacated properties for valuable components like copper wiring and brass fixtures. Once these materials are extracted, criminal scrappers tend to rescind territorial rights to the property and move onto the next foraging area; meanwhile, other criminal scrappers acting as scroungers collect remnant metal in the property upon discovery. However, not all scrappers extracting from previously-scrapped properties are scroungers; for instance, otherwise difficult-to-reach and unwieldy support beams, piping and wiring in opened vacant buildings can be reached by producers through further demolition using powered handsaws, sledgehammers, blowtorches, explosive M-80 firecrackers, or arson—which can be allegedly contracted in Detroit for as little as five dollars (M. Chohaney, interview, July 11, 2014). The latter scenario is evident at the expansive abandoned Packard Plant, which has been mined for decades; yet, producers continue to use dismantling techniques to expose otherwise hidden metal resources—particularly steel support beams. This practice of returning to previously scrapped sites with innovative techniques to extract more, otherwise unattainable resources resembles the hydrofracturing or "fracking" techniques currently practiced by the oil and gas industry.

Efficient mining strategies are essential to maximize the amount of extracted metal. Strategies improve with experience and advice from seasoned scrappers; however, the speed at which metal sources (resource patches) are produced and exploited was frequently cited as the true determinant of a scrapper's success. Adequate speed is achieved by continually monitoring properties for indications of vacancy or entry opportunities, initial entry produces a competitive advantage through the ability to conduct a "pioneer survey" for valuable scrap within the property. Insider information regarding neighborhood property statuses may hasten resource patch creation, but often requires bribery unless the scrapper holds close social connections with local residents. Scrounging among criminal scrappers most profitably occurs in large properties, particularly factories, which decrease in marginal value to producers once the most valuable metals are extracted; therefore, large caches of less-valuable metals (e.g. steel rib bars) are available for scroungers to mine on a first-come basis (Winterhalder and Lu 1997, 49). Mainstream scrappers gain a competitive advantage by planning routes according to residential garbage pickup times and regularly revisiting lucrative areas.

Scrappers compete for resources, but reciprocity-based sharing occurs among scrapper's of rapport. Mainstream and criminal scrappers noted they contact other scrappers when they have too much material to transport and cannot guard the remaining resources. Sharing behavior is motivated by the expectation of reciprocation or information on new sites, and costs little to the producer because the extra scrap would likely be otherwise stolen is left unattended; however, resource theft is often not taken lightly—responses range from social exclusion (a bad reputation) to physical assault (M. Chohaney, field notes, September 25, 2014). Additionally, small groups (usually pairs) of scrappers frequently cooperate with one another to share transportation or handle large pieces or quantities of scrap, a situation termed "conditional mutualism" within the social foraging literature meaning both cooperators simultaneously benefit from working together (Giraldeau and Caraco 2000, 14). Conditional mutualism is most commonly manifested when vehicle-owning scrappers employ those without access to a personal automobile to extract and carry materials to the vehicle, or simply guard the vehicle or act as a lookout. Sometimes the vehicle owner shares less than half of the earnings with the rider, considering the extra profit an overhead charge for fuel costs and vehicle access (M. Chohaney, field notes, September 12, 2014). Most criminal scrappers work in crews of two; one person assists in entering the property and watches the truck and/or neighborhood, and the other associate extracts the metal—a third member may be employed to assist if scrap is abundant enough to economically justify additional profit sharing (M. Chohaney, field notes, September 12, 2014; Le Roux et al. 2009, 1105).

While cooperation among scrappers exists, criminal scrappers are much more concerned with minimizing attention from law enforcement. Scrapping is avoided in areas rendered too risky to mine, typically affluent neighborhoods or locations where scrappers have been acting particularly brazen. One Detroit scrapper, "J," recounted an instance when another scrapper began stealing garage doors from occupied homes, attracting police to the area; consequently, J and the other criminal scrappers operating in that neighborhood could no longer extract resources. Once the identity of the rogue was made certain, J and about five other scrappers physically assaulted the individual, reportedly resolving the issue (M. Chohaney, interview, July 10, 2014).

8.15 Foraging Approaches

Foraging distributions are determined by the perceived earning potential and predation risk of an area (Folmer et al. 2012, 1). In the context of scrapping, predators are law enforcement, disgruntled residents, and fellow scrappers apt to steal. As stated before, Detroit's criminal scrappers frequently task one crew member as a lookout for police or other predators. Maintaining friendly relations with immediate neighbors reduces predation risk from vigilant citizens and law enforcement. While thieving scrappers and concerned residents are major concerns, criminal scrappers' vigilance is usually directed towards avoiding law enforcement because of the high cost incurred if arrested and convicted (Robinette and Ha 2001, 450–1). The distribution of mainstream scrappers is primarily determined by contractual agreements with construction or demolition businesses, or simply the quality of residential waste materials; however, residents may turn against scrappers if they become disruptive and haphazardly mine refuse piles.

Territoriality among criminal scrappers is temporary; properties or sections of large properties are defended from intrusion during the mining process by overt threats of violence and then discarded once the desired value is extracted (M. Chohaney, field notes, July 22, 2014). The nature of criminal scrapping is not conducive to territoriality because many properties are rapidly mined every day, requiring a fairly wide geographic range to meet self-ascribed mining resource thresholds. Thus, defending areas or properties already stripped of valuable materials is unprofitable because guarding a relatively worthless resource comes at the opportunity cost of not mining resource-rich properties.

8.16 Resource Predictability

The spatial predictability of social foragers' resources depends on their scrapping strategy, namely traplining, the act of repeatedly visiting areas where resources have been successfully extracted. Traplining is more productive than relying on social information or random searches when resources are fairly abundant; in fact, Deygout et al. (2010) found traplining approaches yield more resources than relying on social information to identify resources (the social information method) when foragers had over a 20 % chance to find a resource within a given area. This corresponds well with all scrappers' tendency mine areas or properties they have reasonable amount of certainty to find valuable metal.

Mainstream scrappers use a "hard" traplining technique, entailing repeated visits to reliable sites. Roadside refuse piles are replenished weekly and apartment complexes are periodically rich resource patches during common moving seasons. Similarly, sub-contracting with demolition or construction contractors can also be viewed as hard traplining because solicitation by scrappers is motivated by access to reliable resource patches.

Criminal scrappers appear to employ a mild traplining approach, involving revisiting resources patches while using personal information to continually randomly sample the environment for more resources (Deygout et al. 2010, 1137). Mild traplining is less sensitive to spatial predictability than hard traplining, but is potentially inefficient because it relies on variable personal information (i.e. experience) and, to a lesser extent, social information, such as word-of-mouth or shadowing and mimicking foraging strategies of other scrappers (Deygout et al. 2010, 1137). By continually seeking and successfully finding new properties, criminal scrappers protect themselves from crowding and competition while maintaining profitability.

8.17 Choosing Mining Locations and Durations

Social foragers exit a resource patch when the amount of extractable energy, this case scrap metal, falls below the mean uptake (mining) rate (Hancock and Milner-Gulland 2006, 2095). The mean uptake rate is essentially the optimal value consistently obtained with full knowledge of alternative patches. Searching for new patches is less efficient than exiting a productive patch before the extraction rate diminishes to the mean (Hancock and Milner-Gulland 2006, 2099).

Detroit contains an abundance of scrap metal patches; therefore, the distribution of scrappers approaches the ideal free distribution, meaning the number of scrappers is proportional to the amount of extractable metal in the region. However, the ideal free distribution is not realistically applicable to scrappers because they have incomplete patch information, varying risk-taking behavior, and unequal access privileges. Foragers with greater and timely personal knowledge about the location of lucrative patches are generally more successful; for example, advanced information about untapped properties is crucial for a criminal scrapper's intake rate and awareness of lucrative neighborhoods and move-in dates impact the efficiency of mainstream scrappers (Hancock and Milner- Gulland 2006, 2101). The amount of risk scrappers are willing to take also tends to positively correspond with earnings, particularly among criminal scrappers. For instance, criminal scrapper, "J," fitted a plastic dampener around the motor of a gas powered hand-saw for noise suppression in order to cut metal from vacant homes at night. Additionally, he recounted a coordinated effort with 15 other scrappers to break into an abandoned (but sealed) public school building (M. Chohaney, interview, July 10, 2014). Security guards' patterns were monitored and, at the opportune time, they managed to enter at through an unsealed window in an interior courtyard. The whole building was scrapped and the earnings were allegedly divided evenly, a payout worth approximately $1000 dollars per-scrapper.

Scarcity also determines where social foragers choose to extract resources and opportune times to join other foragers to exploit patches (Ruxton et al. 2005, 857). If recently-opened properties or profitable refuse piles are easy to find, the incentive for choosing heavily-mined structures or piles is minimized. In this case, from a spatial perspective, scrappers are more likely to choose sources closer to their position unless distant patches have intake rates high enough to justify the travel cost. Detroit's multitude of vacant properties and the metro area's large population (14th largest in the United States) provide enough resource patches for scrapper density to remain relatively low, reducing the incentive for intruding. As resource patches become scarce, scrappers will spend more time extracting relatively lower-value metals, thus lowering the mean intake rate (Ruxton et al. 2005, 860–1). Mainstream scrappers lose revenue when their routes become crowded; however, criminal scrappers are at highest long-term risk for lowered earnings because vacant properties are finite and therefore become scarcer over time. Furthermore, police surveillance and neighborhood resistance to scrapping cause inconsistent short-term access to resource patches.

8.18 Lessons from Scrappers

Scrappers are archetypical examples of urban miners and naturally exhibit traits of social foragers. Mainstream scrappers should be considered a cost-saving resource for urban sustainability efforts that would reduce the need for costly recycling facilities. Criminal scrappers take economical risks to earn extra money without the overhead of municipal permission and regulation; however, easier access to legal mining rights from property owners may turn the city's menace into a valuable asset. Scrappers could be employed as formal urban miners if planners could guarantee a sufficient mean uptake rate (i.e. average earnings), which could be achieved by mapping and calculating potential resource patch output and optimizing and routing the number of scrappers permitted in an area—similar to the way behavioral ecologists plan feeding grounds for social foragers (see Phipps et al. 2013).

A weighted-overlay site selection analysis may be the most feasible way to estimate the profitability of urban mining resource patches. A tentative example measuring the potential bountifulness of criminal scrapper resource patches can be found in Fig. 8.6; additionally, data sources and the weighting scheme are disclosed in Table 8.4. The proportional weights are based on Chohaney's field-informed perception of factors that informal criminal scrappers consider when identifying resource patches, including abundance (the quantity of unoccupied properties), predation risk (proximity to nearest police station/high population density) and resource quality (older homes). Future classifications should be directly informed by a sur-

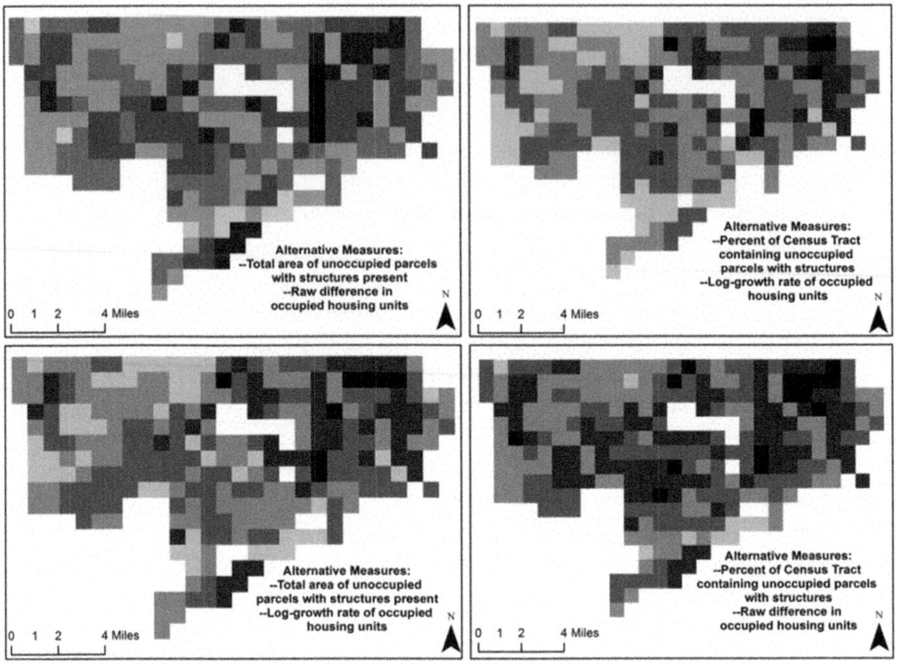

Fig. 8.6 Site selection informal scrapper resource patches In Detroit city. *Darker color gradients* represent more lucrative areas

Table 8.4 Variables, weights, and data sources for criminal scrapper resource patch site selection

Variable	Weight (%)	Data source
Total area of unoccupied parcels with structures present	30	Motor City Mapping's Detroit Residential Parcel Survey (2014)
Percent of Census Tract containing unoccupied parcels with structures	30	Motor City Mapping's Detroit Residential Parcel Survey (2014)
Raw difference in occupied housing units, 2000–2012	20	2007–2012 ACS; 2000 Census
Log-growth rate of occupied housing units, 2000–2012	20	2007–2012 ACS; 2000 Census
Population density	15	2007–2012 ACS
Centroid distance from nearest police station	10	Custom dataset produced by geocoding police stations in Detroit city and metropolitan area
Percent of the population living in poverty	10	2007–2012 ACS
Median age of homes	5	2007–2012 ACS
Median home value	10	2007–2012 ACS

vey of informal criminal scrappers, akin to a participatory GIS site selection approach similar to Mekonnen and Gorsevski (2015).

All four classifications indicate the city's northeast side, near-west side, and northwest corner (including the embattled Brightmoor neighborhood) are the most potentially lucrative extraction areas. This result corroborates with the perception of most of the interviewed scrappers, but needs extensive ground- truthing for validation. Admittedly, this analysis does not consider the amount of scrap metal already extracted from ideal areas and is confounded by the fact criminal scrappers work at a property-level scale individual properties and do not necessarily seeks areas with particular locational attributes; however, the most ideal properties likely lie within locations with relatively dense supplies of feasibly-obtained and profitable minable properties.

Scrappers' harsh and dangerous working conditions are an issue and should be addressed by freely issuing safety equipment (e.g. dust masks or hardhats). Widespread working condition improvements will likely promote greater legal acceptance and regulation, similar to scrapper cooperatives in the developing world that successfully organized and advocated for fairer scrap-buying markets (Medina 2000). Realistically, however, regulation and formal organization of informal scrapping would likely compromise informal scrappers' resource availability, profit margins and mobility. Nevertheless, further research is needed to adequately determine whether informal scrapping can be regulated and harnessed as means to legal and constructive urban sustainability in Detroit and other shrinking cities.

8.19 Conclusion

The United States is a major node in the world's multi-billion dollar scrap metal recycling industry; therefore, its supply chain is worthy of research. Scrapping is a sustainable urban-based industry facilitating energy-saving materials loops involving the collection and processing of scrap metal from informal and industrial sources, which is then processed and sold to metal producers who trade their fabricated metals back to manufacturers. The latter was validated in Detroit, as industrial firms tended to locate near scrap yards, presumably to utilize the efficiencies theorized in the industrial symbiosis literature.

While prompt scrap is the largest source of scrap metal, a significant portion of total supply originates from informal scrappers who furtively mine urban landscapes for dispersed and oftentimes difficult-to-extract discarded metals. Despite the Detroit's varying distribution of discarded metal sources, informal urban mining in and around the city is by no means a random or disorganized economic niche; in fact, Detroit's scrappers have adopted specific, risk-dependent, and sometime illegal metal collection approaches analogous to other social foragers. Scrappers are exemplary urban miners; however, disseminating their skills and fortitude among the general public may be challenging because the local scarcity and volatile market prices determine the viability and appeal of urban mining over time and space.

Nevertheless, the formal scrap industry will likely continue to flourish as the world's appetite for metal products grows along with its conscience about the environmental cost of primary metal production.

This prolegomenon to spatially-integrated research on scrap metal recycling as a significant component of urban sustainability in the United States provided several promising study topics, ranging from an ethnographic approach to study the economic strategies of informal scrappers (i.e. urban miners) to the empirical impact of metal prices on green job growth in the metal recycling industry. The authors hope this exploration of potential research themes will inspire others to extrapolate upon the suggested theoretical frameworks and methods, or at least foster curiosity about one of the United States' most dynamic, economically significant, and critically understudied industries.

References

Ahlbrandt RS, Fruehan R, Giarratani F (1996) The renaissance of American steel: lessons for managers in competitive industries. Oxford University Press, New York

Alvarez JVL, Larrucea MA, Quero-C SF, Del Valle AJ (2008) Optimizing the collection of used paper from small business through GIS techniques: the Leganes case (Madrid, Spain). Waste Manag 28:282–293

Association of American Railroads (2014) Freight railroads in Michigan. Retrieved from: https://www.aar.org/Style%20Library/railroads_and_states/dist/data/pdf/Michigan%202012.pdf. Last accessed 15 Oct 2014

Aylen J, Albertson K (2006) Markets in ferrous scrap for steelmaking. Ironmak Steelmak 33:203–212

Blake P, Chamberlain CS, Chambers D, Clarke B, Mendham J (1987) Heavy metal contamination near scrap metal yards and car breakers yards. Sci Total Environ 59:9–18

Borel-Saladin JM, Turok IN (2013) The green economy: incremental change or transformation? Environ Policy Gov 23:209–220

Broto VC, Allen A, Rapoport E (2012) Interdisciplinary perspectives on urban metabolism. J Ind Ecol 16:851–861. doi:10.1111/j.1530-9290.2012.00556.x

Brunner PH (2011) Urban mining: a contribution to reindustrializing the city. J Ind Ecol 15:339–341. doi:10.1111/j.1530-9290.2011.00345.x

Cai W, Wang C, Chen J, Wang S (2011) Green economy and green jobs: myth or reality? The case of China's power generation sector. Energy 36:5994–6003. doi:10.1016/j.energy.2011.08.016

Cembekli A, Bayraktar A (2014) Turkey leads world scrap metal imports by far. *Today's Zaman*, 5 July. Retrieved from: http://www.todayszaman.com/anasayfa_turkey-leads-world-scrap-metal-imports-by-far_352174.html. Last accessed 12 Oct 2014

Chertow MR (2000) Industrial symbiosis: literature and taxonomy. Annu Rev Energy Environ 25:313–337

Chertow MR (2007) "Uncovering" industrial symbiosis. J Ind Ecol 11:11–30. doi:10.1162/jiec.2007.1110

Congressional Research Service (2008) Minerals price increases and volatility: causes and consequences (CRS report to congress, order code RL34695). Stephen Cooney [Coordinator], Washington, DC. Retrieved from: http://www.dtic.mil/dtic/tr/fulltext/u2/a488765.pdf

Cruz C (2014) City light leader fell for copper con. The Seattle Times, 23 June. Retrieved from: http://seattletimes.com/html/localnews/2023913779_carrascocopperxml.html.. Last accessed 20 Oct 2014

Damuth RJ (2011) Estimating the price elasticity of ferrous scrap supply. Nathan Associates. Retrieved from https://www.isri.org/docs/default-source/recycling-analysis-(reports-studies)/estimating-the-price-of-elasticity-of-ferrous-scrap-supply.pdf

Desrochers P (2002) Industrial ecology and the rediscovery of inter-firm recycling linkages: historical evidence and policy implications. Ind Corp Chang 11:1031–1057

Desrochers P, Lam K (2007) Business as usual in the industrial age: (relatively) lean, green and eco-efficient. Electron J Sustain Dev 1:35–46

Desrochers P, Leppälä S (2010) Industrial symbiosis: old wine in recycled bottles? Some perspective from the history of economic and geographical thought. Int Reg Sci Rev 33:338–361. doi:10.1177/0160017610375441

Detropia. Dirs Ewing H, Grady R, Loki Films (2012). Film

Deygout C, Gault A, Duriez O, Sarrazin F, Bessa-Gomes C (2010) Impact of food predictability on social facilitation by foraging scavengers. Behav Ecol 21:1131–1139. doi:10.1093/beheco/arq120

Ehrenfeld J, Gertler N (1997) Industrial ecology in practice: the evolution of interdependence at kalundborg. J Ind Ecol 1:67–79. doi:10.1162/jiec.1997.1.1.67

Environmental Protection Agency (EPA) (2009) Compilation of air pollutant emission factors In: Enyard E (ed) Chapter 12: metallurgical industry, AP 42, 5th edn, vol I. EPA, Washington, DC

Eynard E, Santangelo M, Tabasso M (2014) Geovisualization tool on urban quality: interactive tool for urban planning. TeMA J Land Use Mob Environ (Special issue):365–375

Fahlman BD (2008) Materials chemistry. Springer, Dordrecht

Folmer EO, Olff H, Piersma T (2012) The spatial distribution of flocking foragers: disentangling the effects of food availability, interference and conspecific attraction by means of spatial autoregressive modeling. Oikos 121:551–561. doi:10.1111/j.1600-0706.2011.19739.x

Giraldeau LA, Caraco T (2000) Social foraging theory. Princeton University Press, Princeton

Goodwin ME (1993) Scrap metal scam targeted; gypsies take metal while posing as Native Americans. American Metal Market. Retrieved from Business Insights: Essentials, Web, 14 October 2014, 7 December

Gregson N, Crang M, Ahamed F, Akter N, Ferdous R, Foisal S, Hudson R (2012) Territorial agglomeration and industrial symbiosis: Sitakunda-Bhatiary, Bangladesh, as a secondary processing complex. Econ Geogr 88:37–58

Gruver G, Giarratani F (2005) Modeling geographic ferrous scrap markets: regional prices and interregional transactions in the United States. J Reg Sci 45:313–341. doi:10.1111/j.0022-4146.2005.00373.x

Halcom C (2014) Scrappers cost Detroit more than money: school days, treated water, manpower also casualties. Crain's Detroit Business, 14 March. Retrieved from: http://www.crainsdetroit.com/article/20140314/BLOG011/140319906/scrappers-cost-detroit-more-than-money-school-days-treated-water. Last accessed 22 Oct 2014

Hancock PA, Milner-Gulland EJ (2006) Optimal movement strategies for social foragers in unpredictable environments. Ecology 87:2094–2102

HNTB Corporation (2011) Michigan state rail plan, Technical memorandum #2: existing conditions. Michigan Department of Transportation, Lansing. Retrieved from: http://www.michigan.gov/documents/mdot/MDOT_MI_SRP_Existing_Conditions_Report_FINAL_2-7-2011_1_344865_7.pdf

Horne SJ (2011) Comments submitted by the institute of scrap recycling industries, Inc. to proposed rule amending the definition of solid waste. Retrieved from: http://www.isri.org/docs/default-source/environment/detailed-comments47AC64213BFA7DED0C8771C0.pdf

Institute of Scrap Recycling Industries (ISRI) (2013) Economic impact study, U.S.-based scrap recycling industry: executive summary. John Dunham and Associates, New York. Retrieved from: http://www.isri.org/docs/default-source/recycling-analysis-(reports- studies)/economic-impact-study-u-s-based-scrap-recycling-industry-2013.pdf

Institute of Scrap Recycling Industries (ISRI) (n.d.) Iron and steel scrap. Retrieved from: http://www.isri.org/recycling-industry/commodities-specifications/ferrous-scrap#.VD9Q4vl4qXM. Last accessed 26 Sept 2014

International Labor Organization (ILO) (1997) The iron and steel workforce of the twenty-first century: report for discussion at the tripartite meeting on the iron and steel workforce of the twenty-first century: what it will be like and how it will work. ILO Sectoral Activities Programme, Geneva
Iritani E (2005) Entrepreneur is shipping scrap metal by the box. Los Angeles Times, 23 June. Retrieved from: http://articles.latimes.com/2005/jun/23/business/fi-shipping23
Jackson C (2014) Man found dead in Detroit manhole. Click on Detroit, 24 July. Retrieved from: http://www.clickondetroit.com/news/body-found-in-manhole-in-detroit/27123200
Janssen MA, Van den Bergh JCJM, Van Beukering PJH, Hoekstra R (2001) Changing industrial metabolism: methods for analysis. Popul Environ 23:139–156
Jolly JL (2013) The U.S. copper-base scrap industry and its by-products: an overview, 13th edn. Received from: http://www.copper.org/publications/pub_list/pdf/scrap_report.pdf
Kaiser MJ (2008) A review of ship breaking and rig scrapping in the Gulf of Mexico. Ocean Dev Int Law 39:178–199. doi:10.1080/00908320802013701
King AJ, Narraway C, Hodgson L, Weatherill A, Sommer V, Sumner S (2011) Performance of human groups in social foraging: the role of communication in consensus decision making. Biol Lett 7:237–40. doi:10.1098/rsbl.2010.0808
Koba M (2013) Copper theft 'like an epidemic' sweeping US. CNBC, 30 July. Retrieved from http://www.cnbc.com/id/100917758. Last accessed 14 Apr 2015
Kooi BR (2010) Theft of scrap metal (Guide No. 58). Center for Problem-Oriented Policing. Retrieved from: http://www.popcenter.org/problems/metal_theft/print/
Le Roux A, Cherry MI, Gygax, Manser MB (2009) Vigilance behaviour and fitness consequences: comparing a solitary foraging and an obligate group-foraging mammal. Behav Ecol Sociobiol 63:1097–1107. doi:10.1007/s00265-009-0762-1
Lyons D, Rice M, Wachal R (2009) Circuits of scrap: closed loop industrial ecosystems and the geography of US international recyclable material flows 1995–2005. Geogr J 175:286–300. doi:10.1111/j.1475-4959.2009.00341.x
Mann O, Kiflawi M (2014) Social foraging with partial (public) information. J Theor Biol 359:112–119. doi:10.1016/j.jtbi.2014.05.044
Marley M (2007) Bigger isn't always better in the US ferrous scrap export game. American Metal Market, 19 January. Retrieved from: https://www.advancedsteel.com/assets/press/American-Metal-Market-Jan-2007.pdf
Marshall A (1986 [1920]) Principles of economics, 8th edn. Macmillan, London
Masocha M (2006) Informal waste harvesting in Victoria falls town, Zimbabwe: socio- economic benefits. Habitat 30:838–848. doi:10.1016/j.habitatint.2005.09.007
Medina M (2000) Scavenger cooperatives in Asia and Latin America. Resour Conserv Recycl 31:51–69
Mekonnen AD, Gorsevski PV (2015) A web-based participatory GIS (PGIS) for offshore wind farm suitability within Lake Erie, Ohio. Renew Sustain Energy Rev 41:162–177. doi:10.1016/j.rser.2014.08.030
Neavling S (2014) 9 ways scrappers are destroying Detroit area. Motor City Muckraker, 11 April. Retrieved from: http://motorcitymuckraker.com/2014/04/11/9-ways-scrappers-are-destroying--detroit-area/1/. Last accessed 20 Oct 2014
Norgate T, Haque N (2010) Energy and greenhouse gas impacts of mining and mineral processing operations. J Clean Prod 18:266–274
Occupational Health & Safety Administration (OSHA) (n.d.) Green job hazards: waste management and recycling: scrap metal recycling–lead. Retrieved from: https://www.osha.gov/dep/greenjobs/recycling_scrap_metal_lead.html. Last accessed 15 Oct 2014
Oosting J (2014) Michigan seeks to crack down on metal theft with new law signed by Gov. Rick Snyder. MLive, 10 April. Retrieved from: http://www.mlive.com/lansing-news/index.ssf/2014/04/michigan_looks_to_crack_down_o.html. Last accessed 17 Oct 2014
Ostrolenk B (1941) Economic geography. Richard D. Erwin, Chicago

Parker B (2000) Brave new shipping world. Scrap Magazine, May/June. Retrieved from: http://www.isri.org/news-publications/scrap-magazine/all-scrap-articles/brave-new-shipping-world#.VG1OHvnF9lw

Phipps WL, Willis SG, Wolter K, Naidoo V (2013) Foraging ranges of immature African white-backed vultures (Gyps africanus) and their use of protected areas in southern Africa. PLoS ONE 8:1–11. doi:10.1371/journal.pone.0052813

Rada EC, Ragazzi M, Fedrizzi P (2013) Web-GIS oriented systems viability for municipal solid waste selective collection optimization in developed and transient economies. Waste Manag 33:785–792

Ravilious K (2013) Mid-town miners: the hunt for urban treasure. New Scientist. Retrieved from: http://www.newscientist.com/article/mg21829192.500-midtown-miners-the-hunt-for-urban-treasure.html

Robinette RL, Ha JC (2001) Social and ecological factors influencing vigilance by northwestern crows, Corvus caurinus. Anim Behav 62:447–452. doi:10.1006/anbe.2001.177

Ruxton GD, Fraser C, Broomb M (2005) An evolutionarily stable joining policy for group foragers. Behav Ecol 16:856–864

Scrap Metal Regulatory Act. Mich. Stat. §§ 5–6 (2014)

Shi H, Chertow M, Song Y (2010) Developing country experience with eco-industrial parks: a case study of the Tianjin Economic-Technological Development Area in China. J Clean Prod 18:191–199. doi:10.1016/j.jclepro.2009.10.002

Sidebottom A, Belur J, Bowers K, Tomson L, Johnson S (2011) Theft in price-volatile markets: on the relationship between copper price and copper theft. J Res Crime Delinq 48:396–418

Söderholm P, Ejdemo T (2008) Steel scrap markets in Europe and the USA. Miner Energy 23:57–73. doi:10.1080/14041040802018497

Stessel RI (1996) Recycling and resource recovery engineering: principles of waste processing. Springer, Berlin

Stuart EJ (2013) Implications of the minimill era. [Web blog], 24 April. Retrieved from: http://www.steelorbis.com/blog/2013/04/24/implications-of-the-minimill-era/. Last accessed 20 Oct 2014

Sumner S, King AJ (2011) Actions speak louder than words in socially foraging human groups. Commun Integr Biol 4:755–757

United Nations (2005) The millennium development goals report. Retrieved from: http://unstats.un.org/unsd/mi/pdf/mdg%20book.pdf

United Nations Environment Programme (UNEP) (2008) Green jobs: towards decent work in a sustainable, low-carbon world. Worldwatch Institute, Nairobi. Retrieved from http://www.ilo.org/wcmsp5/groups/public/---ed_emp/--- emp_ent/documents/publication/wcms_158727.pdf

Winterhalder B, Lu F (1997) A forager-resource population ecology model and implications for indigenous conservation. Conserv Biol 11:1354–1364. doi:10.1046/j.1523-1739.1997.95184.x

Winterhalder B, Smith EA, American Anthropological Association (1981) Hunter-gatherer foraging strategies: ethnographic and archeological analyses. University of Chicago Press, Chicago

Wolfe M (2011) Detroit scrap city. In these times, 8 June. Retrieved from: http://inthesetimes.com/article/11478/detroit_scrap_city

Wolman A (1965) The metabolism of cities. Sci Am 213:179–190

Xiarchos IM (2005) Steel: price links between primary and scrap markets. Paper presented at the Southern Agricultural Economics Association Annual Meetings: Little Rock, Arkansas, February. Retrieved from: http://ageconsearch.umn.edu/bitstream/35655/1/sp05xi01.pdf

Xiarchos IM, Fletcher JJ (2009) Price and volatility transmission between primary and scrap metal markets. Resour Conserv Recycl 53:664–673

Yin CY, Ghazaly SM, Hilmi BM (2007) Chemical stabilization of scrap metal yard contaminated soil using ordinary Portland cement: strength and leachability aspects. Build Environ 42:794–802

Chapter 9
Where Are the Garden(er)s? Examining Gardener Motivations and Community Garden Participation-Sheds in Austin, Texas

Ronald R. Hagelman III, Gregory S. Mast, and Colleen C. Hiner

Abstract Community gardens have received increasing attention from a wide range of academics, professionals, activists, hobbyists, students, and politicians as potential solutions to problems as diverse as food insecurity, childhood obesity, social fragmentation, economic instability, and declining biodiversity. Community gardens serve as sources of food and nutrition in addition to playing a role in cultural, political, economic, and ecological systems at multiple scales. As such, much work has been done to catalogue the benefits of community gardens to participants and society at large. However, less is known about how the benefits of community gardens translate into individual motivations to participate in community gardens or the relative strength of these motivating factors in terms of inspiring gardeners to overcome impediments to participation such as distance to access. Furthermore, although the inherently spatial focus of the discourse surrounding local food networks has led to efforts to map the food systems of both urban and rural areas at the state, census tract, and even neighborhood level, there is a deficit in research designed to visualize these patterns. Using a mixed methods approach, this article addresses these deficits in current research by assessing the motivations for participation among community gardeners and visualizing, using geographic information systems (GIS), the geographic participation-sheds of a sample of community gardens within the Austin, Texas, metropolitan area. The results of this inquiry inform scholarly discourses on the evolving role of community gardens within urban space. Moreover, through modeling the spatial patterns of community garden participation-sheds and characterizing the motivations of participating gardeners, this study serves as a guide to urban land managers as they seek to develop comprehensive and sustainable land management policies.

Keywords Urban gardens • Sustainable agriculture

R.R. Hagelman III (✉) • C.C. Hiner
Department of Geography, Texas State University, San Marcos, TX, USA
e-mail: rhagelman@txstate.edu; cchiner@txstate.edu

G.S. Mast
Community Forester, Tree Folks, Austin, TX, USA
e-mail: greg.mast@gmail.com

J.D. Gatrell et al. (eds.), *Urban Sustainability: Policy and Praxis*,
Geotechnologies and the Environment 14, DOI 10.1007/978-3-319-26218-5_9

9.1 Introduction

Although community gardens have existed for many decades, urban managers have embraced and incentivized them in recent years as a mechanism for addressing social, environmental, and economic sustainability goals. As a result, community gardens have received increasing attention from a wide range of academics, professionals, activists, hobbyists, students, and politicians. They have been proposed as potential solutions to problems as diverse as food insecurity, childhood obesity, social fragmentation, economic instability, and declining biodiversity. Academics in particular have focused on community gardens not only as sources of food and nutrition, but also their role in cultural, political, economic, and ecological systems at multiple scales (Smith and Kurtz 2003; Baker 2004; Glover 2004; Beilin and Hunter 2011; Evers and Hodgson 2011).

Though much work has been done cataloguing the benefits of community gardens to participants and society, what is not as well understood is how these benefits translate into expressed motivations for participation in community gardens or the spatial patterns that result from the participation of those motivated to engage in community gardening. Both would be of potential value to a wide range of proponents of community gardening and urban managers interested in incorporating them into planning and development approaches focused on improving urban sustainability.

Although the inherently spatial focus of the discourse surrounding local food networks has led to efforts to map the food systems of both urban and rural areas at the state, census tract and even neighborhood level, there is a deficit in research designed to visualize these networks at the scale of individual behaviors (Hu et al. 2011; Hubley 2011; Kremer and DeLiberty 2011; Russell and Heidkamp 2011). The purpose of this research is to qualitatively explore the motivations of community gardeners in Austin, Texas as well as to quantitatively model the spatial patterns of their participation using a geographic information system (GIS) to support cartographic visualization of community gardener participation-sheds in Austin, Texas. The results of this inquiry promise to inform scholarly discourses on the evolving role of community gardens within urban space as well as inform urban land managers of the characteristics of community garden participation-sheds and the motivations of participating gardeners.

9.2 Literature Review

Numerous authors have argued the negative social, environmental, public health, economic, and security consequences of both the industrialization of food production and its absence from the consciousness of most eaters (Carson 1962; Merrill 1976; Schlosser 2001; Nestle 2002; Pollan 2006). Some of the responses to these assertions have been the creation of 'local' and 'alternative' food systems within and adjacent to cities. Organic, biointensive, ecological, and permaculture approaches to food production; as well as community gardening, co-operative marketing outlets for farmers, and urban agriculture are examples of some of these emergent responses (Kaufman and Bailkey 2000; Jeavons 2006; Ingram 2007; Nordahl 2009). One of the oldest of these movements, community gardening, can trace its roots back to the late nineteenth and early twentieth century with the promotion of urban food gardens by the mayor of Detroit and the passing of the Allotment Acts in Britain (Glover 2003; Pudup 2008; Milburn and Vail 2010). Until the late 1970s, community gardens were largely temporary in nature and intent and conceived of as responses to the problem of a momentarily insufficient food supply (e.g. Liberty and Victory Gardens during the World Wars, Relief Gardens of the Great Depression, and Potato Patches of the 1890s). During the 1970s, community garden projects began to prioritize concerns such as neighborhood beautification, building social capital, and enhancing community ties (Smith and Kurtz 2003; Pudup 2008). Community gardening in the twenty-first century is characterized by an enormous diversity of forms, locations, intentions, and strategies. While Pudup's (2008) "three tributary discourses and movements evincing collective resistance and individual self-improvement that, taken together, animate contemporary organized garden projects" (pg. 1232) focuses on urban renewal, the therapeutic benefits of plants, and urban ecology, other authors include concerns about nutrition and food security, maintenance of cultural traditions, social protest, and the economic impact of gardens (Airriess and Clawson 1994; Patel 1996; Baker 2004; Saldivar-Tanaka and Krasny 2004; Cutter-Mackenzie 2009; Evers and Hodgson 2011; Litt et al. 2011). Collectively, these themes define the potential motivations explored by this study and inform the qualitative investigations of Austin's community gardeners.

Following recent efforts to map local food systems, our research also offers cartographic visualizations of the spatial patterns of Austin community gardeners' participation. Concerns about the energy consumption associated with transporting food long distances and the economic impact of globalization on local communities has led to calls for a reinvigoration of local and alternative food systems. In response, there has been a push to understand the geographic extent of the networks of production and consumption that provide cities with food, also referred to as 'foodsheds' (Hu et al. 2011). Kremer and DeLiberty (2011) visualize the local foodshed of Philadelphia, including urban farms, community gardens, institutional gardens, urban orchards, farmers markets, restaurants, institutions, co-ops, and food retail in the analysis. Our research adds to this body of work by offering visualizations of community garden participation-sheds. A second area of research has been the identification of food deserts. These efforts range from simple 'distance to supermarket' and income ranges for census tracts, to more complex models using service area networks for various retail outlets integrating variable distance thresholds for different locations (Hubley 2011; Russell and Heidkamp 2011; United States Department of Agriculture Economic Research Service 2011). Modeling food access, however, is very challenging as it is different for each person and place (Hubley 2011). Thus far, work to pair foodshed analysis with food desert mapping seems to be lacking. In part, this may result from the fact that the existence of food deserts is contested in some academic and government publications (i.e. Ver Ploeg et al. 2009). Regardless of specific definitions, efforts to visualize local, urban, and regional food production, distribution, and access have been greatly enabled in recent years with the application of GIS and computer cartography. In addition to the qualitative investigation of gardeners' motivations, our research offers both a novel method for collecting the spatial data necessary to illustrate neighborhood-scale geographic patterns of community gardener participation, or participation-sheds, and cartographic visualizations of a sample of Austin's current community gardeners.

9.3 Methods and Data

Austin, the capitol city of Texas, has a population of approximately 800,000 people, with an economy based on government, education, high tech industry, health care, and business services (Austin Chamber of Commerce 2011). Community gardening in Austin extends back at least as far as 1978, the year the oldest currently operating community garden was formed. There are anywhere from 26 to 32+ community gardens in Austin, though the exact number of community gardens depends on how community garden is defined. The Coalition of Austin Community Gardens maintains a listing of some 32 garden sites with a total of more than 465 plots, though again this number is approximate due to the variation in how the different gardens are run. For example, some gardens have individual plots for individual gardeners, while others maintain only large commonly cultivated plots (Coalition of Austin Community Gardens 2012). In 2011, the City of Austin hired a 'Sustainable Urban Agriculture Coordinator', tasked in part with the coordination of community gardens throughout the city. The creation of this position has occurred roughly concurrently with the construction of new community gardens, legal recognition and definitions for community gardens and urban farms, and alternative permitting for water use on urban farms and vegetable gardens (City of Austin 2011).

Although there is much discussion in the academic literature on the precise definition of community gardens (e.g. Blake and Cloutier-Fisher 2009; Firth et al. 2011; Turner et al. 2011), the City of Austin has adopted a specific legal definition for a 'city supported community garden'. According to city code, a city supported community garden "is used by a group of four or more participating gardeners either on separate plots or farmed collectively by the group to grow, produce and harvest food crops for personal or group use, consumption or donation by the non-profit organization or cooperatively for the benefit of its members" (City of Austin 2011). This definition has the advantage of being locally defined, specific to food production and exclusive of entrepreneurial urban farming. It does not, however, make any distinction between gardens that are collectively farmed versus gardens that are made up of separate plots. In reality, many gardens have both, but those gardens for which individual plots are the main focus are in the majority. Indeed, of roughly 32 gardens in a regional community gardens list maintained by the Coalition of Austin Community Gardens, just 4 gardens are reported as having only collectively managed growing areas (Coalition of Austin Community Gardens 2012). On the other hand, there are at least 465 individual plots across Austin's other community gardens, with each plot representing at least 1 gardener. Owing to their great preponderance, gardeners of individual plot gardens are the focus of this research.

Table 9.1 Selected Austin community gardens

Garden name	Location	Plots	Founded
Alamo Community Garden	2101 Alamo St.	29	1994
Deep Eddy Community Garden	3001/2 Atlanta	34	1978
Festival Beach Community Garden	35 Waller St.	80	2010
Sunshine Community Gardens	4814 Sunshine Drive	150	Early 1970s
Blackshear Community Garden	2011 E. 9th St.	8	2008

In an effort to standardize our gardener population to some degree, we have excluded gardens sponsored by or located on the grounds of overtly ideological groups, such as religious organizations, and gardens associated with outside communities (e.g. Boy Scout troops, schools, social clubs, residential co-operatives). Instead, this research will focus on gardens for which the organizing commonality of membership is judged to be proximity to the garden, and which are located on publicly owned or accessible land. This approach focused our analysis on gardeners associated with five community gardens in Austin, Texas (Table 9.1). These gardens together contain a total of 336 plots, and represent a selection of the more established, actively tended gardens around the city. They also all contain reference to a particular street, neighborhood or park in their names and are distributed throughout the city in both more and less economically affluent neighborhoods.

This research employed a mixed methods approach and has been determined to be exempt from full or expedited review by the Texas State Institutional Review Board. Field data were collected during June, July, and August of 2011. Distances and patterns of travel were established by asking gardeners for the closest street intersection to their home, generating a relatively accurate yet anonymous geocoding of gardeners' points of origin. Although this method of situating individual survey respondents relies heavily on the veracity of their responses, it generates a data file with a high degree of precision and, to our knowledge, has not previously been employed in GIS analysis. To identify spatial patterns in gardener participation, geocoded gardener origins were paired with garden locations to construct participation sheds for each garden using ArcGIS v10.0 Geographic Information System (GIS). Following Kremer and Deliberty (2011), network distance was determined using a shortest path analysis for each gardener-garden pair and aggregated to determine mean distance traveled. Visualizations of selected participation-sheds (Figs. 9.1, 9.2, 9.3, 9.4, and 9.5) are discussed below.

Semi-structured interviews were conducted with gardeners including questions designed to prompt gardeners to provide explanations for why they participate in the community garden, whether and how those reasons have changed with time, which of their reasons for participating are strongest, and why and how they decided to participate in their particular community garden. The interview responses were recorded on a digital voice recorder and manually coded by the author to identify and describe motivational themes. During the coding process, particular emphasis was placed on the themes outlined in the literature above, including Environmental Quality, Relaxation and Recreation, Psychospiritual Benefits, Nutrition and Food, Cultural Identity, Social Interaction, and Economics. These themes are discussed in the qualitative results section.

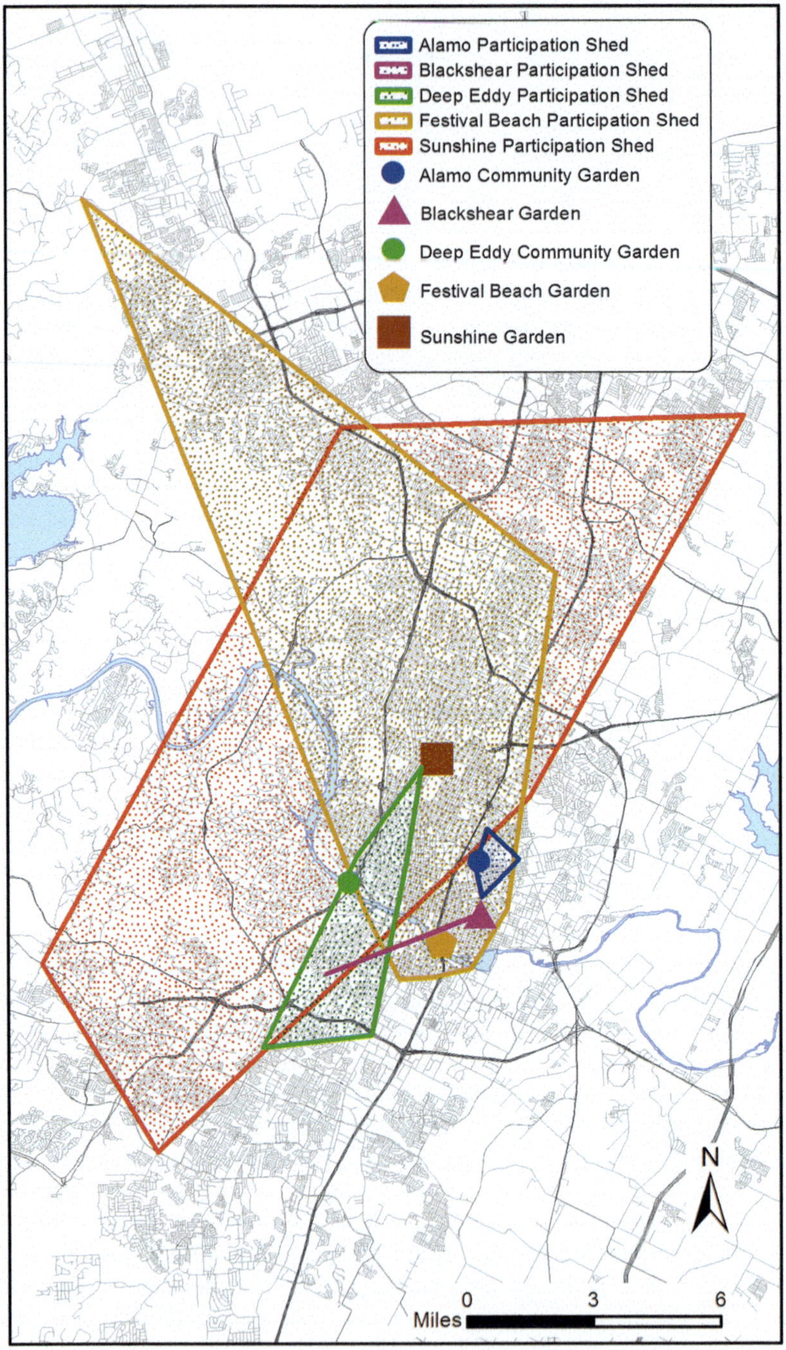

Fig. 9.1 Selected Austin Community Garden participation-sheds

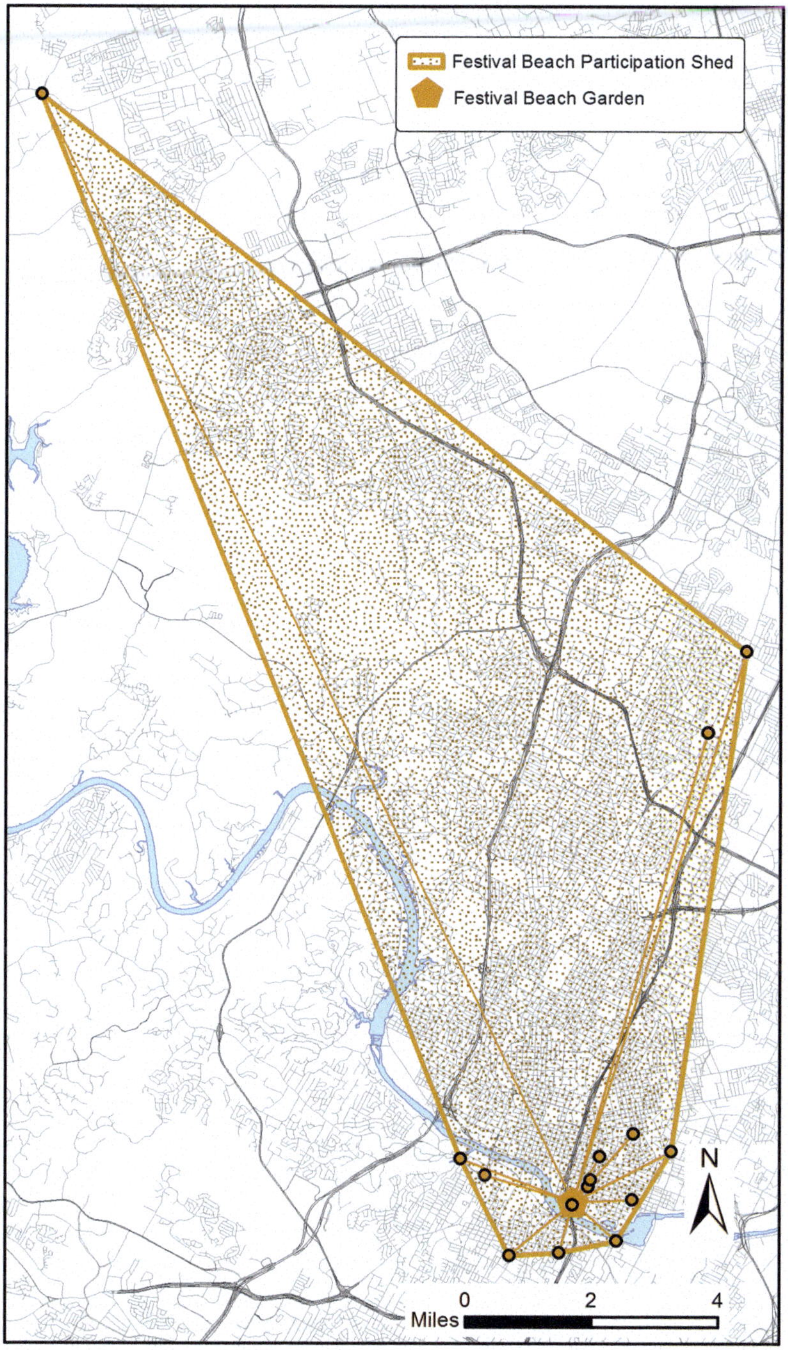

Fig. 9.2 Festival Beach Community Garden

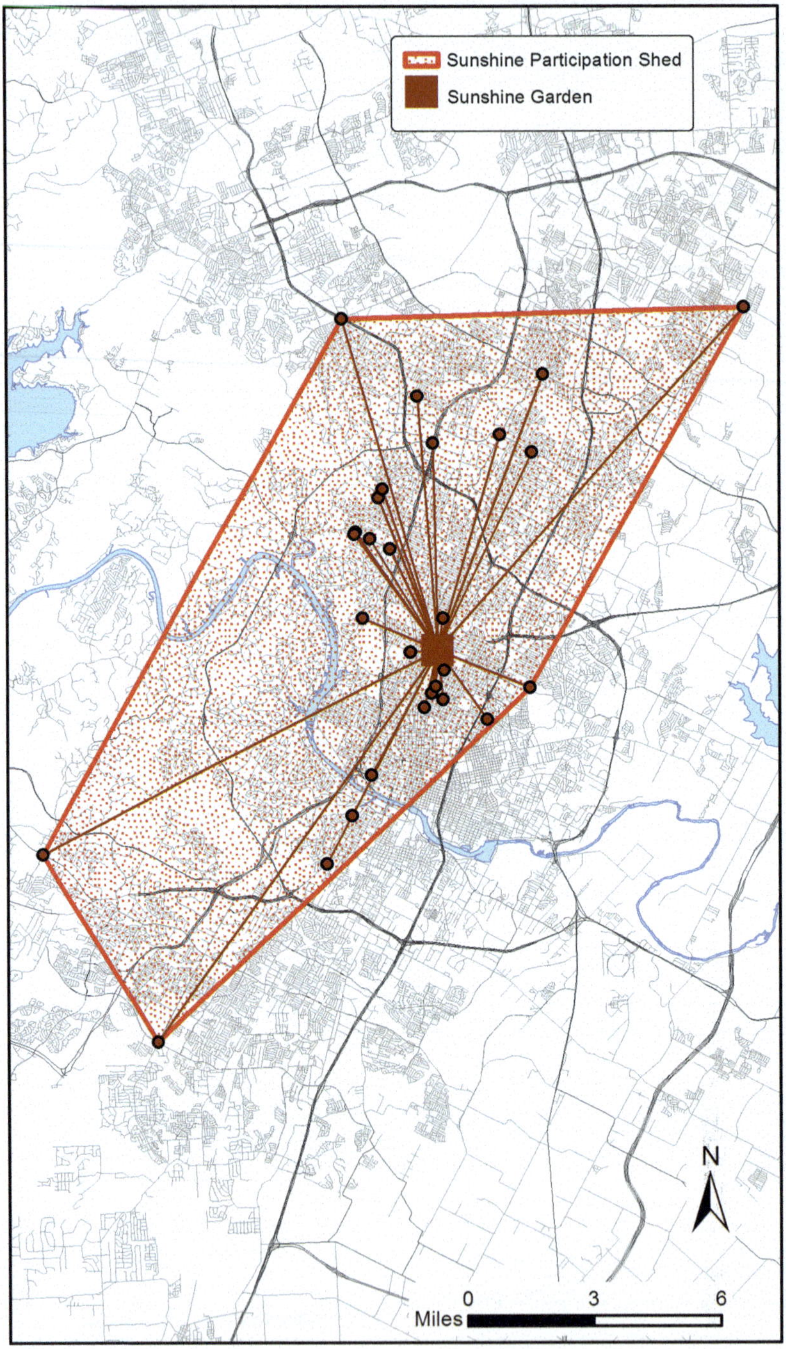

Fig. 9.3 Sunshine Community Garden

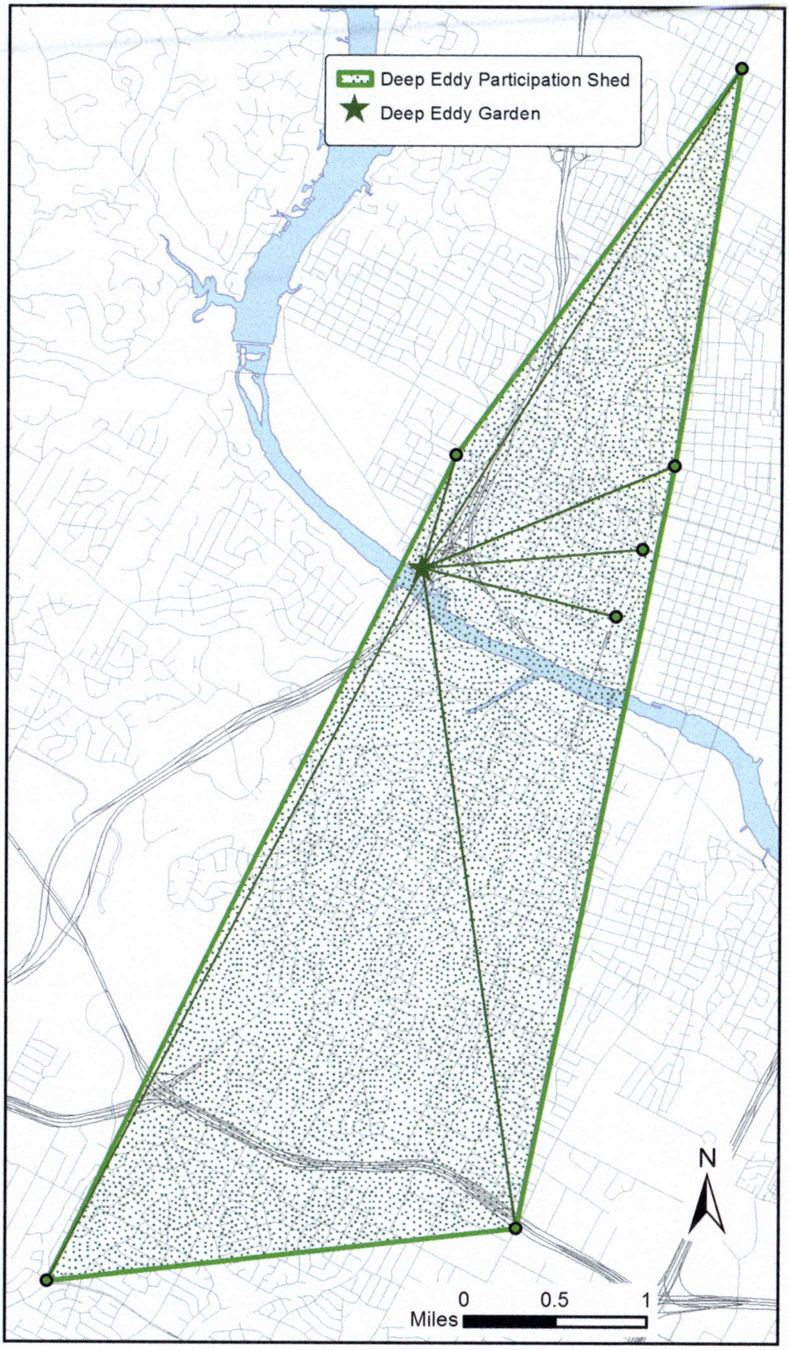

Fig. 9.4 Deep Eddy Community Garden

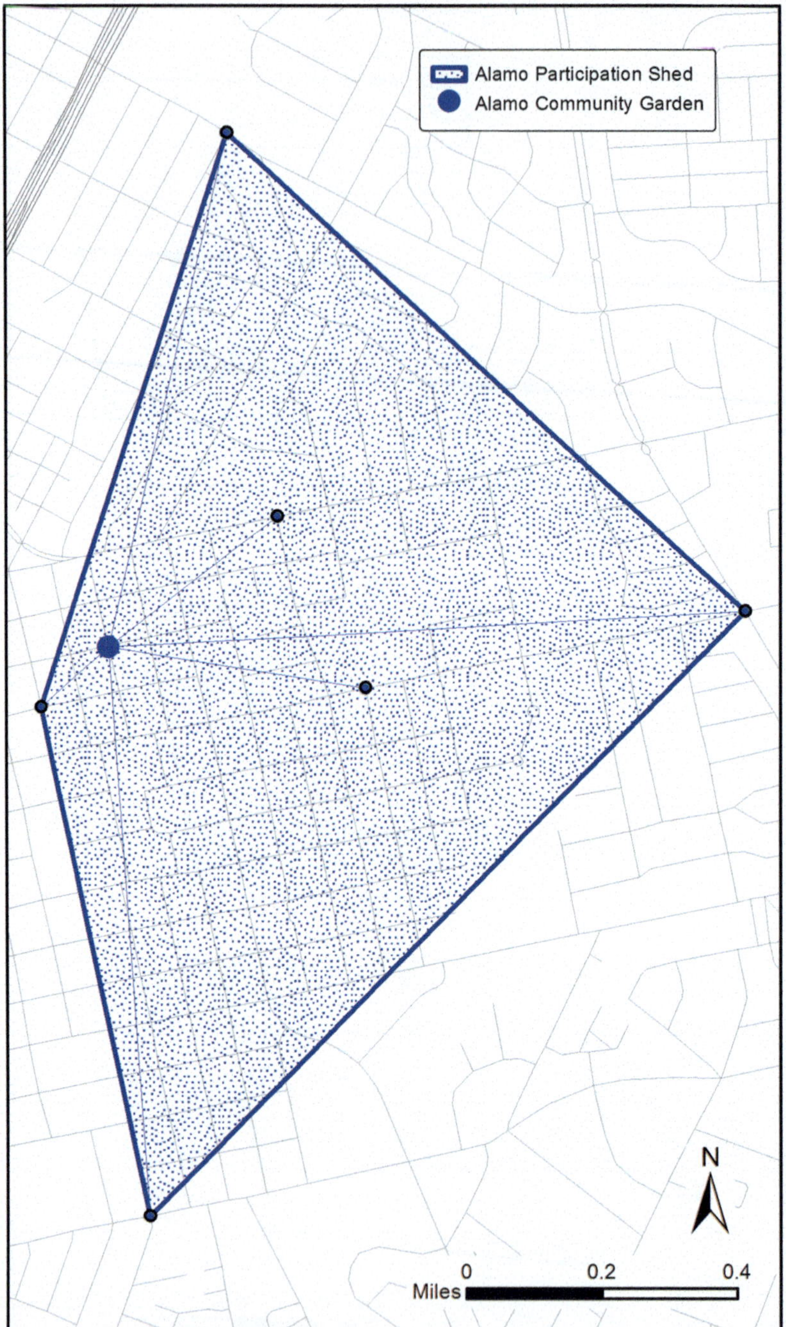

Fig. 9.5 Alamo Community Garden

9.4 Quantitative Results and Visualizations

In total, 63 ($n=63$) community gardeners were surveyed and interviewed for this study. The majority were female (59 %) and, among all gardeners, most possessed a bachelors or master degree (68 %), tended to live in detached/single-family housing (68 %), tended to use a personal car for transportation to and from the garden (70 %), and reported a mean age of 48.3 years. Table 9.2 provides additional information on the time spent gardening, years of experience, and distance and travel time. Overall, gardeners in this study tended to be experienced, committed to an average of nearly 5 h per week of gardening, and willing to travel over 6 miles (mean of 10 min) to visit their garden site.

Utilizing the entire data set ($n=63$), participation-sheds for five community gardens (Fig. 9.1) were determined according to the GIS methods described above. Most gardens are centrally located and associated with either the University of Texas campus area, central-east Austin, or the recent suburban areas of southwest and northwest Austin. Two gardens, Festival Beach Community Garden (Fig. 9.2) and Sunshine Community Garden (Fig. 9.3), have participation-sheds that range across much of the Austin metro region.

These two gardens represent the oldest (Sunshine, est. early 1970s) and the newest (Festival Beach, est. 2010) of the gardens. In the case of Festival Beach, the larger participation-shed largely results from a single outlier in northwest Austin. Without that data point, the participation-shed would be more concentrated in central Austin. From the interviews with the gardeners at Sunshine, it was clear that a number of them had started working in this garden at a time when they were living closer to the garden (usually related to education or employment at the University of Texas in central Austin) and have subsequently moved to one of the rapidly expanding suburbs around Austin. This result is of potential interest to garden managers and city planners interested in sustainability and who seek to engage and retain younger gardeners who may eventually find themselves moving to suburban settings as they age and gain affluence or find themselves priced out of central locations as a result of gentrifications trends.

Figures 9.4 and 9.5 illustrate Deep Eddy Community Garden and Alamo Community Garden, respectively. These gardens exhibit neighborhood-scale participation-sheds. The Blackshear participation-shed (see Fig. 9.1) should be considered in relation to the meager data available, but if the trend is consistent among its gardeners, it does not indicate a wide-ranging participation-shed. Deep Eddy

Table 9.2 Selected characteristics of community gardeners

	N	μ	σ
Hours/week	63	4.92	5.12
Years of experience	63	18.1	15.3
Distance (km)	63	6.27	7.05
Travel time (min)	63	9.14	7.22

Community Garden (Fig. 9.4), also one of the longer standing gardens established in 1978, shows signs of the central-to-suburban migrations of its gardeners as well. This is supported by onsite interviews. The Alamo Community Garden (Fig. 9.5) is the one in our sample group that exhibits a true neighborhood-scale participation-shed. This garden is situated in east Austin and in the heart of a neighborhood that is undergoing a rapid gentrification from Hispanic and African American to young, mostly white and affluent, professionals. This entire region of Austin is being remade to suit the expectations of the new residents and the tight spatial participation of the community garden reflects this purposeful place-making trend (City of Austin 2003; KTBC News 2013).

Although this analysis would benefit from a larger data set, it does illustrate the utility of both the anonymous gardener-origin data collection technique and the value of visualizing participation sheds on a garden-by-garden basis. First, precise spatial data related to individual gardeners can be collected in this manner without having to manage confidentiality. This technique can be used for other data collection goals as well, but in this case it helped build a rapport with the gardeners that was quick, non-intrusive, and allayed concerns about being identified or contacted beyond their survey responses. Secondly, the GIS-based visualizations are useful in identifying unique spatial and participatory characteristics of different gardens and rendered information that can be useful to city planners and garden managers across any metropolitan area. Underlying these spatial patterns are individual and group motivations for participation in community gardening. These are analyzed for our study area using qualitative methods in the section below.

9.5 Qualitative Results

Although the geographic patterns of participation discussed above can tell us many things about the overall characteristics of individual gardens, choosing to engage in community gardening is a personal decision that is often related to individual goals (i.e. health, recreation, environmental concerns, etc.). In this section we describe the motivational themes exhibited among our sample group, including Environmental Quality, Relaxation and Recreation, Psychospiritual Benefits, Nutrition and Food, Cultural Identity, Social Interaction, Economics, and Access to Land and Resources.

9.5.1 Environmental Quality

Concern for the environment among gardeners was chiefly associated with a desire to consume "food that [didn't] have to [be] transport[ed] too much" (Festival Beach) and avoidance of agrochemicals, considered to be "bad for the environment" (Sunshine). Although some participants at Festival Beach and Sunshine Garden mentioned more localized environmental benefits such as bird seed plots and wildlife habitat areas, concern for the direct ecosystem services of garden spaces was not a strong motivation for gardeners as a whole. Environmental education, though arguably the avenue through which urban community gardens have the greatest potential to make a positive environmental impact, was mentioned only once as a motivating factor. This is not to say that a great deal of environmental education does not occur in community gardens, only that gardeners did not identify it as a significant motivating factor.

9.5.2 Relaxation and Recreation

The assertion that gardening can be a relaxing and enjoyable recreational pursuit is well supported by the responses of community gardeners in this study. Community gardens serve as places where people can go outside and unwind or simply enjoy the pleasant surroundings. These outdoor recreational spaces serve as a way to "de-gas" (Alamo) after a stressful day in the office, spent sitting "in front of a computer" (Sunshine). Community gardens provide knowledge workers an opportunity to create something tangible and immediate, in contrast to the work seen as "90 % in my head" (Sunshine).

Although, as one gardener put it, "[y]ou can buy vegetables at the grocery store" (Sunshine), (community) gardening produces something in addition to fresh local produce: an opportunity to relax and relieve stress. According to respondents, the relaxation and stress-relieving effects of gardening supersedes the utility of the produce itself. In other words, the practicality and productiveness of gardening is what makes the activity attractive and relaxing to gardeners. Moreover, this practicality and productivity is what differentiates gardening from a sport or game in which recreation is the only objective. Community gardening provides an opportunity to "put in two hours and be able to see an immediate difference" (Sunshine).

9.5.3 Psychospiritual Benefits

Psychospiritual motivations include the therapeutic benefits of both gardening and natural environments, as well as the role of gardens as sacred spaces. The healing, "antidepressant" (Sunshine) properties of "literally standing on dirt" (Sunshine) was mentioned repeatedly by gardeners. (Community) gardens serve as places where people who feel disconnected from material reality can go to reconnect with their bodies through creative action. Through the embodied practice of gardening, gardeners are linked corporeally and emotionally to not only the produce, but also to the activity of gardening, their fellow gardeners, and to the space and place of the garden. For some gardeners, this connection facilitates prayer and feeling "closer to God than a lot of people get at church" (Sunshine).

9.5.4 Nutrition and Food

Gardeners were drawn to the perceived purity of garden produce, the health benefits of eating fresh vegetables, and the superior taste relative to store bought produce. Because of its freshness and their role in it production, gardeners felt confident that the food they grow does not need "15 preservatives" (Blackshear) in order to be edible, and can be eaten without fear of "chemicals…salmonella, [and] e-coli" (Deep Eddy). Concern for health and nutrition was thus oriented more towards rejecting the undesirable aspects of conventionally grown or purchased food in favor of garden produce, rather than on adding fruits and vegetables to a diet otherwise lacking in such foods. At the same time, community gardeners were more concerned with the superior taste of fresh garden produce, especially tomatoes, than the direct health benefits of the food. After being "picked a week or two weeks" (Deep Eddy) prior to purchase, grocery store produce is found "lacking" (Sunshine) in quality.

9.5.5 Cultural Identity

As "all human activity is culturally encompassed" (Johnston et al. 1994: pg. 116), in essense, all gardeners are motivated based on the values of their culture. However, some gardeners specifically mentioned family or ethnic tradition as important in the decision to join a community garden. Austin is at the heart of a rapidly growing urban area, with people arriving from all over the United States and from other countries. Some gardeners, including those who "grew up in rural areas" (Sunshine) and whose "family…were all farmers" (Alamo), were drawn to community gardens as a way of "getting some of [their] own" (Sunshine) land in the middle of a dense urban area, "far away from…family" (Alamo). To these respondents, community gardening was a mechanism to continue family and ethnic traditions of community service at the local and personal scale, accomplished by giving away produce to neighbors, for example (Festival Beach and Alamo).

9.5.6 Social Interaction

Community gardens are places where people go to build and maintain relationships, express their political ideals, and learn from one another. Gardeners give produce to friends, family and coworkers, cementing those relationships, and undertake communal projects to improve the garden and the surrounding communities. Although people are often initially inspired to join a community garden "because [a] friend has a plot already" (Deep Eddy), continued participation is supported by "feel[ing] part of a community [and] establishing roots somewhere" (Alamo). As creative spaces, community gardens also provide an opportunity for people to enact alternative political visions, such as a city in which "everyone could be within walking or biking distance to a farm" (Blackshear). Alternatively, and in contrast to community gardens' role as communal spaces, community gardens also draw some participants in search of "private space and private time" (Sunshine) away from the responsibilities of work and family life.

Because different gardeners have different motivations, "community [interests]...are sometimes in contradiction" (Alamo) with the interests of individual gardeners. Much of the variation between community gardens is rooted in varying approaches to mediating that conflict, with some gardens organized along strict adherence to formalized rules and bylaws, others driven largely through consensus decision making, and still others at the direction of a charismatic individual.

9.5.7 Economics

Direct economic motivations, such as being "able to save money gardening (Sunshine)" were very rare among gardeners, with most comments made in the context of describing their evolving level of skill at gardening rather than expressing an economic motive. The hours of labor and inputs required to produce large crops of vegetables present a relatively large opportunity cost compared to produce from the supermarket or farmers' market. Only two gardeners expressed a direct economic motivation for participating, one of which was doing so in preparation for retirement when she would "have less income [and] need to grow some food" (Sunshine), but not out of immediate necessity.

9.5.8 Access to Land and Resources

Aside from social interaction, what distinguishes community gardens from private gardens is the opportunity to share tools, space, and knowledge within a group. Not having "access to gardens for a long time" (Sunshine) and experiencing "concrete shock" (Alamo) after moving into a cramped urban apartment can provide the impetus for joining a community garden, but being able to "trust the soil and [the] space" (Alamo) sustains involvement over the long haul. The superior growing conditions in established gardens provide "a better place to garden than at home" (Deep Eddy). Other gardeners see community gardens as social resources, where they can rely on others to "take care of the plants" (Blackshear) when on vacation. The physical proximity of other gardeners creates an environment where sharing and cooperation are the norm. Gardeners at every garden mentioned the opportunity to ask plot neighbors for advice as a primary motivation for being involved with the garden. These communities of knowledge are thus spatially constructed and place-bound.

9.5.9 *The Value of Community Gardening, in Sum*

Drawing from findings related to each of these themes – Environmental Quality, Relaxation and Recreation, Psychospiritual Benefits, Nutrition and Food, Cultural Identity, Social Interaction, Economics, and Access to Land and Resources – it is clear that community gardening holds significant value for gardeners and is a means for building stronger, more cohesive, and ultimately more self-sustaining communities. In addition to serving as spaces for education, environmental improvement, and food production, (community) gardens also serve as places for social interaction and solitude, spiritual renewal, cultural enrichment, and community building. In sum, community gardens have the potential to fulfill multiple aims at once and, thus, could be an ideal mechanism for managers and citizens to enact sustainability at the local level.

9.6 Conclusions

The purpose of this research was to explore the spatial patterns, or participation-sheds, and motivational themes of community gardeners in Austin, Texas. In order to accomplish this we gathered point of origin data from a sample of gardeners, generated geographic visualizations of participation among five community gardens in Austin, and surveyed and interviewed 63 gardeners regarding their experience, background, garden-related travel habits, and personal motivations for participating in their respective community gardens. In relation to the geographic visualizations, two types of participation-sheds were observed. In most cases, gardeners were willing to travel distances that rendered city-wide (regional) participation-sheds. Qualitative results show that the reason for the larger participation-sheds vary, but were often related to changing lifestyles among gardeners who remain loyal to their initial garden of choice, but find themselves farther flung within the urban area due to lifestyle migration.

Although additional research and larger sample sizes should be pursued, this analysis illustrates the spatial and motivational complexities underlying community garden participation. If our results are indicative of community gardeners overall, and urban planners seek to expand the many benefits of local food production and community development, successful establishment of new gardens may not be as simple as situating them in low-value, open lots while ignoring the complex socio-cultural context of urban landscapes. In a neighborhood like east Austin where the inertia for current community development is clearly place-based, this approach may work. But, given the propensity for public community gardens to be situated in urban cores, the majority of our results indicate that many gardeners, although intending to remain loyal to their original neighborhoods garden, may face the disincentive of ever increasing distance as their lives (and livelihoods) change over time and draw them out to suburban areas. While departing gardeners could theoretically be replaced by newly proximate ones, existing and anticipated shifts in residential patterns illustrate the complexity of meeting both the needs of current and future gardeners while allowing for continuing land development.

Despite the publicity and popularity of local food production in Austin, and the apparent and multiple benefits of community gardening, our sample indicates a sizable percentage of Austin hosts neither public community gardens nor gardeners. Further analysis may reveal that these regions of the city are served by other gardening opportunities (e.g., schools, churches, businesses, etc.), however, if local food production and consumption is a development or planning goal, there may be more that needs to be done to encourage and promote (community) gardening in these regions, both among suburban affluent neighborhoods and economically challenged neighborhoods throughout the city.

References

Airriess CA, Clawson DL (1994) Vietnamese market gardens in New Orleans. Geogr Rev 84(1):16–32

Austin Chamber of Commerce (2011) Business establishments by industry: Greater Austin profile. Homepage of Austin Chamber of Commerce. http://www.austin-chamber.org/. Last accessed 4 Apr 2011

Baker LE (2004) Tending cultural landscapes and food citizenship in Toronto's community gardens. Geogr Rev 94(3):305–325

Beilin R, Hunter A (2011) Co-constructing the sustainable city: how indicators help us "grow" more than just food in community gardens. Local Environ 16:523–538

Blake A, Cloutier-Fisher D (2009) Backyard bounty: exploring the benefits and challenges of backyard garden sharing projects. Local Environ 14(9):797–807

Carson R (1962) Silent spring. Houghton Mifflin, New York

City of Austin (2003) Staff task force on gentrification in East Austin: findings and recommendations. City of Austin, Austin

City of Austin (2011) CHAPTER 14-7. Sustainable urban agriculture. City code. http://www.amlegal.com/nxt/gateway.dll/Texas/austin/title14useofstreetsandpublicproperty/chapter14-7sustainableurbanagriculture?f=templates$fn=default.htm$3.0$vid=amlegal:austin_tx. Last accessed 20 Apr 2012

Coalition of Austin Community Gardens (2012) Gardens. http://communitygardensaustin.org/?page_id=24. Last accessed 20 Apr 2012

Cutter-Mackenzie A (2009) Multicultural school gardens: creating engaging garden spaces in learning about language, culture, and environment. Can J Environ Educ 14:122–135

Evers A, Hodgson NL (2011) Food choices and local food access among Perth's community gardeners. Local Environ 16(6):585–602

Firth C, Maye D, Pearson D (2011) Developing "community" in community gardens. Local Environ 16:555–568

Glover TD (2003) Community garden movement. In: Christensen K, Levinson D (eds) Encyclopedia of community. Sage, Thousand Oaks. http://www.sage-ereference.com.libproxy.txstate.edu/view/community/n103.xml?rskey=UMJx04&result=1&q=community%20garden. Accessed 26 Apr 2012

Glover TD (2004) Social capital in the lived experiences of community gardeners. Leis Sci 26(2):143–162

Hu G, Wang L, Arendt S, Boekenstedt R (2011) Analyzing sustainable, localized food production systems with a systematic optimization model. J Hunger Environ Nutr 6:220–232
Hubley TA (2011) Assessing the proximity of healthy food options and food deserts in a rural area in Maine. Appl Geogr 31(4):1224–1231
Ingram M (2007) Biology and beyond: the science of "Back to Nature" farming in the United States. Ann Assoc Am Geogr 97(2):298–312
Jeavons J (2006) How to grow more vegetables* (and fruits, nuts, berries, grains and other crops) *than you ever thought possible on less land than you can imagine. 10 Speed Press, Berkeley
Johnston RJ, Gregory D, Smith DM (eds) (1994) The dictionary of human geography. Blackwell, Cambridge, MA
Kaufman J, Bailkey M (2000) Farming inside cities: entrepreneurial urban agriculture in the United States. Lincoln Institute of Land Policy, Cambridge
Kremer P, DeLiberty TL (2011) Local food practices and growing potential: mapping the case of Philadelphia. Appl Geogr 31(4):1252–1261
KTBC News (2013) Cost of East Austin transformation into a hip neighborhood. KTBC Fox 7 News. http://www.myfoxaustin.com/story/22212259/cost-of-East-Austin-transformation-into-a-hip-neighborhood. Last accessed 19 Dec 2014
Litt JS, Soobader M-J, Turbin MS, Hale JW, Buchenau M, Marshall JA (2011) The influence of social involvement, neighborhood aesthetics, and community garden participation on fruit and vegetable consumption. Am J Public Health 101(8):1466
Merrill R (ed) (1976) Radical agriculture. New York University Press, New York
Milburn L-AS, Vail BA (2010) Sowing the seeds of success. Landsc J 29(1):71–89
Nestle M (2002) Food politics how the food industry influences nutrition and health. University of California Press, Berkeley
Nordahl D (2009) Public produce: the new urban agriculture. Island Press, Washington, DC
Patel IC (1996) Rutgers urban gardening: a case study in urban agriculture. J Agric Food Inf 3(3):35–46
Pollan M (2006) The omnivore's dilemma: a natural history of four meals. Penguin Press, New York
Pudup MB (2008) It takes a garden: cultivating citizen-subjects in organized garden projects. Geoforum 39(3):1228–1240
Russell SE, Heidkamp CP (2011) 'Food desertification': the loss of a major supermarket in New Haven, Connecticut. Appl Geogr 31:1197–1209
Saldivar-Tanaka L, Krasny ME (2004) Culturing community development, neighborhood open space, and civic agriculture: the case of Latino community gardens in New York City. Agric Hum Values 21(4):399–412
Schlosser E (2001) Fast food nation. Houghton Mifflin, New York
Smith CM, Kurtz HE (2003) Community gardens and politics of scale in New York City. Geogr Rev 93(2):193–212
Turner B, Henryks J, Pearson D (2011) Community gardens: sustainability, health and inclusion in the city. Local Environ 16:489–492
United States Department of Agriculture Economic Research Service (2011) Food desert locator. http://www.ers.usda.gov/Data/FoodDesert/. Last accessed 20 Apr 2012
Ver Ploeg M, Breneman V, Farrigan T, Hamrick K, Hopkins D, Kaufman P, Lin B-H, Nord M, Smith T, Williams R, Kinnison K, Olander C, Singh A, Tuckermanty E (2009) Report summary. Access to affordable and nutritious food—measuring and understanding food deserts and their consequences: report to congress. United States Department of Agriculture Economic Research Service. Administrative Publication No. (AP-036). http://www.ers.usda.gov/Publications/AP/AP036/. Last accessed 20 Apr 2012

Chapter 10
Sustainability Attitudes and Actions: An Examination of Craft Brewers in the United States

Mark W. Patterson, Nancy Hoalst-Pullen, and W. Blake Pierson

Abstract Making beer is a multifaceted endeavor. Arguably, there are over 14 production stages in beer making, which can be simplified down to three stages – the brewing stage, the fermentation and maturation stage and finally, the packaging and distribution stage. Each production stage provides an opportunity for the brewery to engage in [more] sustainable views and practices. Given the focus of this edited volume, we were particularly interested in the economic, environmental and social sustainability attitudes and actions of craft breweries located in urban areas. Specifically, we wanted to explore the similarities and differences in sustainability attitudes and actions among urban craft breweries, as well as different brewery types, categorized here as brewpubs, microbreweries, and regional craft breweries. Using the three pillars of sustainability, we developed a survey designed to ascertain the attitudes and practices embraced by these brewery types. Findings show minor differences among the brewery types under the economic pillar, while brewpubs significantly differed from microbreweries and regional/regional craft breweries under the environmental and social pillars.

Keywords Sustainability • Beer • Craft • Breweries • Brewpubs

10.1 Introduction

Urban breweries
Beer sustainability
One keg at a time

M.W. Patterson (✉) • N. Hoalst-Pullen • W.B. Pierson
Department of Geography and Anthropology, Kennesaw State University, Kennesaw, GA 30144, USA
e-mail: mpatters@kennesaw.edu

J.D. Gatrell et al. (eds.), *Urban Sustainability: Policy and Praxis*, Geotechnologies and the Environment 14, DOI 10.1007/978-3-319-26218-5_10

Craft breweries are defined by three overarching tenets: they are small (produce less than six million US barrels[1] annually), use traditional and/or innovative ingredients,[2] and are independently owned (with 25 % or less economic control or ownership by a non-craft brewer). Of the 3400 breweries currently operating in the United States, nearly 99 % can be classified as craft breweries. As categorized by the Brewer's Association (2015a), craft breweries include brewpubs, microbreweries, regional craft breweries (RCB) and craft brewing companies. Brewpubs are breweries with associated restaurants or bars that sell directly to the consumer, while microbreweries are craft breweries that produce less than 15,000 barrels annually, of which 75 % is sold off-site. Regional craft breweries are breweries that make traditional beers and generally have a larger market and production scale when compared to microbreweries. Contract brewing companies are breweries that are hired for marketing or producing beer. Additional non-craft brewery categories (recognized by the Brewer's Association 2015a) include regional breweries that produce less than six million barrels annually and large breweries that produce more than six million barrels annually. Finally, one emerging craft brewery market segment that is not defined by the Brewer's Association is craft nanobreweries, which are independently owned, small batch breweries. While the size of these nanobreweries is debated, the most common size restriction is a three-barrel (six kegs) maximum, although upwards of ten barrel systems are commonly classified as nanobreweries.[3]

The U.S. craft brewing scene has proliferated since its inception in the late 1970s. In 1980, there were eight craft brewers (Brewer's Association 2015b); by 2009, there were 1596. Five years later, the number of craft breweries more than doubled to 3418 (Brewer's Association 2015c), and included 1412 brewpubs, 1871 microbreweries and 135 regional craft breweries (Brewer's Association 2015c). Currently, craft breweries are found in all 50 states and Washington, D.C. In fact, craft breweries are becoming so commonplace that craftbeer.com states "The average American lives within 10 miles of a brewery."[4]

While U.S. craft beers account for only 11 % of all beer sold, its market share is increasing and threatens (or arguably annoys) the bottom line and declining market share of large-scale conglomerate brewers such as Anheuser-Busch InBev and SABMiller. In response, these international brewing corporations are now producing "crafty" beers like Shock Top and Blue Moon, as well as acquiring US craft breweries such as Goose Island, 10 Barrel, Breckenridge Brewery, and Elysian Brewery. Additionally, many other craft breweries are partially owned by the large international brewing giants. For example, Anheuser-Busch InBev owns 32.2 % of the Craft Beer Alliance Company that is composed of Redhook Ale Brewery,

[1] One barrel equals 31 US gallons, or about 330 12 oz bottles, or 248 16 oz pints.

[2] Traditional refers to the high quality base ingredients (i.e. malted barley).

[3] The authors, however, would define nanobreweries by its production rate, i.e. craft breweries producing less than 1000 barrels per year.

[4] Applied geography conducted by the authors found that each of us lives less than 10 miles from a microbrewery.

Widmer Brothers Brewery, Kona Brewing Company, Omission Beer and Square Mile Cider. So while beer can be defined as craft, "crafty", or corporate, one thing they all have in common is the need for resources – raw ingredients, a lot of water, a lot of energy (e.g. electricity), brewing equipment and space, and brewers[5] to make beer.

10.2 Sustainability in Beer Production

As beer production requires a lot of resources, regardless of the scale of production, there should be an inherent value in incorporating sustainable practices to reduce the use of water and energy, decrease the production of waste and greenhouse gases, implement best practices, and increase social capital for employees, consumers and communities at large.

As a concept, sustainability relates to the balance between current behaviors and conditions and their lasting impact on future economies, societies (politically and culturally), and environments. Popularized by the World Commission on Environment and Development (1987), the concept of the "three pillars of sustainable development" has been adopted by various industries to assess the economic, environmental and social sectors of sustainability. The beer industry – from nanobreweries to large breweries – are looking ever inward and increasingly assessing and improving their own economic viability, environmental responsibilities, and roles in society's constructs and communities, at local to global scales.

For breweries, sustainability practices are interconnected with incoming resources (raw ingredients (grains/malt, hops, yeast) water, energy, packaging, etc.) as well as outgoing resources (beer, air emissions, and solid or liquid wastes). These are part of 14 production stages of beer as identified by Hoalst-Pullen et al. (2014), which begins with the milling of malted grains and ends with the distribution of the final packaged product.[6]

Defining sustainability within the brewing industry is difficult at best. Sustainability in practice at breweries is considered a bit contentious because of the modern-day disconnects between the brewery locations and the raw ingredients (thus lending to de facto increases in the carbon footprints), the innate need for large amounts of water, and the production of liquid or solid wastes at nearly every stage of beer production. In spite of these inherent shortcomings, however, sustainability can be incorporated into the culture of the craft brewery. Brewers and their breweries can and do engage in economic, environmental and/or social sustainability

[5] As production scale increases, the number of workers also does, and expands from the brewer to other positions in such fields as marketing, merchandise, distribution, logistics, human resources, accounting and so on.

[6] Hoalst-Pullen et al. (2014) denoted the beginning of the brewing process as the cracking or crushing of malt into grist. Arguably, we acknowledge that one could start at the true inception of any beer, which is the recipe.

practices that produce innovations which in turn save resources and money, and many times, give back to the employees, consumers and the community at large.

One common example of sustainable practices at breweries is the reduction of water used in producing beer. Indeed, the brewing industry is one of the largest industrial users of water, with water used not just in the brewing process, but in growing grains and hops, in heating, cooling, cleaning and maintaining the brewery, in the pasteurizing, bottling and packaging of beer, and finally, in recycling and waste disposal (Olajire 2012). While beer is more than 90 % water, it is the hidden waters that impact the sustainability of beer, with 3 10 l of effluent (wastewater) produced per 1 l of beer during the brewing process (Kanagachandran and Jayaratne 2006) and a whopping 298 l (78 gal) of water used when including crop production (data derived from Mekonnen and Hoeksra 2010). As noted in a report by WWF, SABMiller and GTZ (2010), the cultivation of crops accounts for over 90 % of total water footprint in beer production.

Using the three pillars of sustainable development, Hoalst-Pullen et al. (2014) developed a survey designed to ascertain which views and praxes are embraced by US breweries. In the study, RCBs were surveyed, and the findings suggest that most are pledging (and frequently succeeding) at implementing various measures related to sustainability, and most are assessing the environmental impacts (and associated economic losses) of their brewery and adopting sustainable practices.

Given the focus of this edited volume on urban sustainability, we were particularly interested in sustainability practices of craft breweries located in urban areas. Within this urban context, we wanted to explore if there were differences in sustainability attitudes and actions among craft brewery types, namely, brewpubs, microbreweries, and regional/regional craft breweries (RCB).[7] Considering the three pillar of sustainability is prevalent in the regional craft beer industry (Hoalst-Pullen et al. 2014), we are interested to compare and contrast sustainability and sustainable practices with craft breweries at all scales of production.

10.3 Methods

We created an online survey instrument to assess sustainability and sustainable practices within US craft breweries. In Summer 2014, a list of 2822 craft breweries (brewpubs, microbreweries and regional craft breweries) was compiled by way of the Brewers Association website. A formal invitation was sent out via email to the craft breweries to participate in the survey. For craft breweries that did not have email addresses, the same information was provided to the brewery by way of

[7] The Brewers' Association (2015a) defines "Regional Craft Breweries" as ones that are independent and producing predominately traditional or innovative beers. They define "Regional Breweries" as ones making between 15,000 and six million barrels. Thus, we combined these two definitions into one category to identify the craft designation as well as size.

contact information webpages on the brewery's website. The craft brewery was not contacted if the contact information (email, phone number, contact information page etc.) did not exist or was inaccurate. We followed up the initial email with two additional emails, each sent 2 months apart, to all breweries that had not responded. Finally, we called and/or sent a fourth and final email to all non-responding breweries 6 months after the initial email.

The online survey consisted of a series of statements regarding the three pillars of sustainability – economic, environmental, and social – as related to the craft beer industry. Respondents (representing their given brewery) were asked on a Likert scale of 1 (strongly disagree) to 4 (strongly agree) whether they agreed with each sustainability-type statement. Further, we asked respondents to indicate various sustainable practices they are (or are not) undertaking at their brewery. Initially, we had them select from the following answers: brewery used to do it, brewery does not it, brewery does it, and brewery plans to do it in the future, where "it" is the action in question. Ultimately, however, we combined the answers into two categories, namely used to do it/do not do it, and currently do it/plan to do it. Finally, respondents could make additional comments at the end of the survey.

We divided the respondents into two groups based on the location of their brewery. Using ArcGIS 10.2.2 we plotted the location of the brewery and overlaid the US Census Urban areas maps to parse out which breweries were located in urban areas. We further subdivided the breweries on the type they self-identified (brewpub, microbrewery, and regional/regional craft brewery.

10.4 Results

A total of 303 breweries respond to our survey (See Fig. 10.1). The mean and median number of years of operation was 9.43 and 5.5 years respectively. The mean and median number of US barrels produced was 17,127 and 1,300. Of the responding breweries, 196 identified themselves as Corporate owned, 59 as Employee owned, 41 as other and 9 gave no answer. Self-classified by craft brewery type, 88 respondents said they represented a brewpub, 132 a microbrewery and 60 a RCB. The remainder did not provide an answer for this question. It should be noted that under the definitions provided by the Brewers Association[8] more than half of our respondents (32) self-identifying as regional craft breweries are regional breweries, as they produced between 15,000 and 6 million barrels annually.

---

[8] http://www.brewersassociation.org/statistics/market-segments/

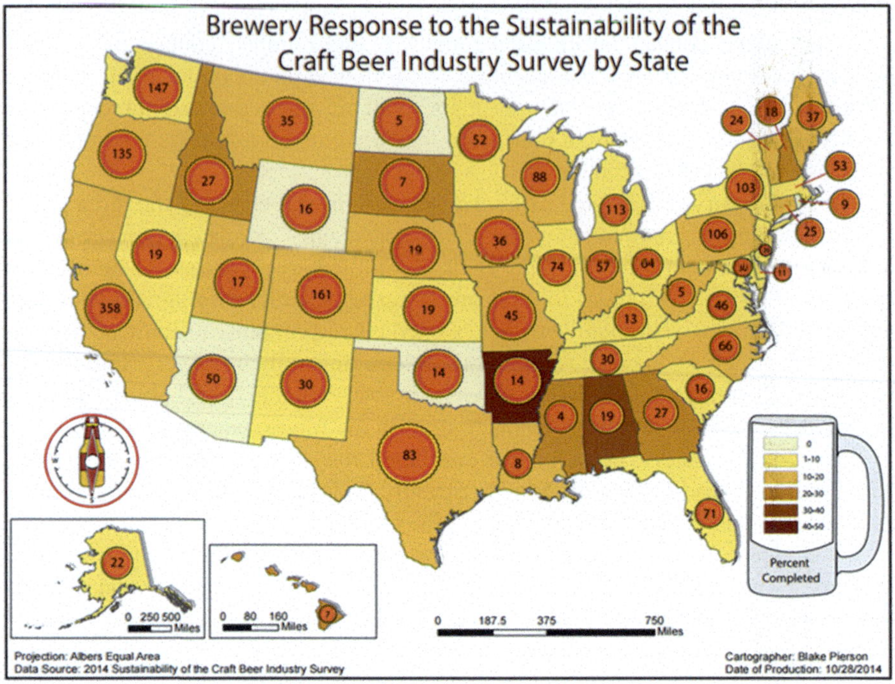

Fig. 10.1 Brewery response to the sustainability of the craft beer industry survey by state

10.4.1 Urban v. Rural Breweries

Of the 303 represented craft breweries, 65 breweries were located in rural areas while 224 were located in urban areas. Fourteen respondents did not provide their brewery name and thus were excluded from analyses and were not mapped.

Urban Brewery Views

Of the 224 urban breweries, 49 were classified as brewpubs, 107 microbreweries and 68 as RCBs. We ran a one-way ANOVA comparing general sustainability responses by brewery type. The results for each statement by craft brewery type are shown in Tables 10.1 and 10.2.

Results from the ANOVA indicate that differences in responses are statistically significant ($p<0.01$) for four statements: (1) using social media to promote sustainability goals and commitments; (2) reducing energy and water using economically justifiably technology; (3) supporting the workforce; and (4) supporting the local community. In addition we ran a Post-hoc Tukey HSD (Honestly Significant Difference) test to determine which brewery type responses were statistically different (Table 10.3).

Table 10.1 Descriptive statistics on urban brewery views

Statement	Type	N	Mean	SD
Brewery uses social media to promote sustainability goals and commitments	Brewpub	68	2.40	1.095
	Microbrewery	107	2.84	0.933
	Regional craft brewery	49	2.88	0.857
Brewery has visible sustainability goals and commitments	Brewpub	68	2.69	0.996
	Microbrewery	107	2.85	0.998
	Regional craft brewery	49	3.06	0.992
Brewery has goals of reducing energy and water uses using economically justifiable technology	Brewpub	68	2.71	1.008
	Microbrewery	106	3.09	0.737
	Regional craft brewery	49	3.27	0.730
Brewery has sustainability goals to use organic or local products and resources	Brewpub	68	2.57	0.982
	Microbrewery	107	2.74	1.049
	Regional craft brewery	49	2.73	0.758
Brewery has sustainability goals of supporting workforce	Brewpub	66	2.91	0.890
	Microbrewery	107	3.13	0.794
	Regional craft brewery	49	3.35	0.729
Brewery has sustainability goals of supporting local community	Brewpub	68	3.47	0.819
	Microbrewery	107	3.68	0.576
	Regional craft brewery	49	3.82	0.391

Table 10.2 ANOVA results for urban brewery views

Statement		Sum of Squares	df	Mean square	F	p value
Brewery uses social media to promote sustainability goals and commitments*	Between groups	9.87	2	4.94	5.25	0.006
	Within groups	207.84	223	0.94		
Brewery has visible sustainability goals and commitments	Between groups	3.90	2	1.95	2.02	0.135
	Within groups	212.94	221	0.95		
Brewery has goals of reducing energy and water uses using economically justifiable technology*	Between groups	10.23	2	5.12	7.47	0.001
	Within groups	150.73	220	0.69		
Brewery has sustainability goals to use organic or local products and resources	Between groups	1.27	2	0.63	0.67	0.512
	Within groups	208.86	221	0.95		
Brewery has sustainability goals of supporting workforce*	Between groups	5.57	2	2.79	4.24	0.010
	Within groups	142.59	217	0.66		
Brewery has sustainability goals of supporting local community*	Between groups	3.65	2	1.83	4.62	0.010
	Within groups	87.48	211	0.39		

*Statistically significant ($p<0.01$)

Table 10.3 Tukey HSD results on general sustainability statements

Statement	Type (I)	Type (J)	Mean diff (I–J)	Std. error	Significance
Brewery uses social media to promote sustainability goals and commitments	Brewpub	Microbrewery	−0.444	0.150	0.010*
		Regional craft	−0.480	0.182	0.024
	Microbrewery	Brewpub	0.444	0.150	0.010*
		Regional craft	−0.036	0.167	0.974
	Regional craft	Brewpub	0.480	0.182	0.024
		Microbrewery	0.036	0.167	0.974
Brewery has goals of reducing energy and water uses using economically justifiable technology	Brewpub	Microbrewery	−0.388	0.129	0.008*
		Regional craft	−0.559	0.155	0.001*
	Microbrewery	Brewpub	0.388	0.129	0.008*
		Regional craft	−0.171	0.143	0.457
	Regional craft	Brewpub	0.559	0.155	0.001*
		Microbrewery	0.171	0.143	0.457
Brewery has sustainability goals of sup-porting workforce	Brewpub	Microbrewery	−0.223	0.127	0.188
		Regional craft	−0.445	0.154	0.012
	Microbrewery	Brewpub	0.223	0.127	0.188
		Regional craft	−0.222	0.141	0.259
	Regional craft	Brewpub	0.445	0.154	0.012
		Microbrewery	0.222	0.141	0.259
Brewery has sustainability goals of sup-porting local community	Brewpub	Microbrewery	−0.212	0.098	0.079
		Regional craft	−0.346	0.118	0.010*
	Microbrewery	Brewpub	0.212	0.098	0.079
		Regional craft	−0.134	0.109	0.434
	Regional craft	Brewpub	0.346	0.118	0.010*
		Microbrewery	0.134	0.109	0.434

*Statistically significant at $p<0.01$

From Tables 10.2 and 10.3, microbreweries (2.84, 0.933) had higher reported use of social media to promote sustainability goals and commitments than brewpubs (2.40, 1.095), $p=0.01$. Likewise, RCBs (2.88, 0.857) had higher uses of social media than brewpubs (2.40, 1.095), $p=0.024$. However, differences between microbreweries and regional craft breweries were not statistically significant. Unlike microbreweries and RCBs, brewpubs offer food in addition to beer and therefore have a walk-in clientele that you see less frequently at microbreweries and RCBs. In fact, microbreweries and RCBs in many states are not permitted to sell beer on the premise,[9] and thus need alternative routes to promote their product. Therefore social media plays a significantly more important role in advertising the beer produced at microbreweries and RCBs, and by default, their sustainable goals and commitments.

[9] Although the loophole for many microbreweries and RCBs is to sell "tours" or related merchandise (serving glasses) and in return, provide "free" beer.

Similarly, both microbreweries (3.09, 0.737), p=0.008 and RCBs (3.27, 0.730) p=0.001 reported higher usage of technology for reducing energy and water use then brewpubs (2.71, 1.008). Again there was no statistically difference between microbreweries and RCBs for this statement. The most obvious reason is that brewpubs use water for activities associated outside of the brewery; namely, the restaurant and/or bar side of the establishment. It is possible that monitoring water and energy usage for just the brewing process is more difficult for brewpubs, and therefore they are less likely to invest in such technologies when compared to microbreweries and RCBs. It could also be an outcome of brewing capacity, as most brewpubs are small (esp. compared to RCBs) and the investment of technologies is minimized in part from the other avenues for profit (e.g. food, non-alcoholic drinks).

The only statistically significant difference in responses for the support of the workforce was between brewpubs (2.91, 0.890) and regional craft breweries (3.35, 0.729), p=0.012. This is most likely a product of the high turnover common in the restaurant business, especially with servers and related kitchen staff. Restaurant job studies (e.g. Ghiselli et al. 2001; Lambert et al. 2001) found connections between turnover intention and job satisfaction, life satisfaction and age, among other demographic factors (e.g. gender, tenure, race etc.). In contrast to the restaurant-oriented brewpubs, it is likely that RCBs consider their employees as longer-term investments with developed skillsets related to production (and/or marketing, distributing etc.). Combined with the larger production scale, RCBs are likely more effective in creating a satisfying work culture for employees surrounding the product(s), the company, or even craft beer industry as a whole.

Finally, statistically significance differences existed only between brewpubs (3.47, 0.819) and regional craft breweries (3.82, 0.391), p=0.01 on supporting the local community. Again RCBs are more likely to engage actively with the local community as a method to increase their public presence. RCBs are likely large enough to have a marketing extension to their company, and supporting the local community coincides with an avenue of advertising their product for public consumption.

10.4.2 The Three Pillars of Sustainability

Economic Sustainability

We started by examining actions undertaken by breweries by positing statements centered on economic and corporate sustainability. The questions and results are shown in Table 10.4.

Regarding statements related to environmental regulations, sustainability pledges, company vision statements and using sustainability as a marketing tool, the chi-squared test results indicate that there is no statistical significance in the differences among the brewery types. This is encouraging as most breweries (95 %) not only meet environmental regulations, but are making commitments to increase

Table 10.4 Economic sustainability Chi squared test results by brewery type

	Type	Used to or does not	Currently does or has plan to	Pearson Chi-square	df	Sig
Does brewery take steps to meet or exceed current environmental regulations?	Brewpub	5	58	3.782	2	0.151
	Microbrewery	6	98			
	Regional	0	49			
Does brewery meet any pledged sustainability commitments?	Brewpub	6	57	1.738	2	0.419
	Microbrewery	13	83			
	Regional	3	43			
Does brewery have a mission/vision statement that includes sustainability commitment?	Brewpub	7	57	3.550	2	0.169
	Microbrewery	11	94			
	Regional	1	48			
Does the brewery uses sustainability as a marketing tool?	Brewpub	26	38	2.311	2	0.315
	Microbrewery	39	66			
	Regional	13	35			
Brewery has a systematic review of operations to assess and improve overall sustainability?*	Brewpub	22	42	8.809	2	0.010
	Microbrewery	26	77			
	Regional	5	44			

*Chi-squared significant $p<0.01$

sustainability. While the differences among brewery types were not statistically significant regarding using sustainability as a marketing tool, 66 % of the respondents reported using sustainability for marketing.

A difference does emerge when asked about a systematic operations review to improve sustainability. Such reviews indicate how breweries are doing regarding sustainability and how and where they can improve. Here, the differences in answers are statistically significant as 52 % of brewpubs and 34 % of microbreweries do not conduct such a review. These percentages seem to contradict the mission/vision statement which includes a commitment to sustainability. Eighty-nine percent and 90 % of the brewpub and microbreweries respectively, claimed to have this commitment in their mission/vision statement.

Environmental Sustainability

Under the pillar of environmental sustainability, we provided respondents with a series of statements concerning their actions. Table 10.5 displays the environmental sustainability responses and analyses.

Six environmental sustainable actions were found to have no statistical difference in responses regarding the brewery type. This means all craft brewery types are behaving in a similar manner on these actions (or at least are reporting similar actions). Regarding the collection and repurposing of spent grains, slightly less than

Table 10.5 Environmental sustainability Chi squared test results by brewery type

	Type	Used to or does not	Currently does or has plan to	Pearson Chi-square	df	Sig
Collect spent grains for other purposes	Brewpub	1	66	0.900	2	0.638
	Microbrewery	2	105			
	Regional	0	49			
Collect and reuse yeast from fermentation for other purposes*	Brewpub	40	26	9.223	2	0.010
	Microbrewery	64	40			
	Regional	18	31			
Reduce packaging materials	Brewpub	13	49	3.621	2	0.164
	Microbrewery	14	89			
	Regional	4	44			
Invest in reusable/recyclable packaging	Brewpub	13	49	3.823	2	0.148
	Microbrewery	14	91			
	Regional	4	45			
Installed energy meters to measure and control consumption*	Brewpub	48	17	27.914	2	0.000
	Microbrewery	44	56			
	Regional	11	34			
Recover heat (examples: from wort cooling, keg water systems)	Brewpub	17	48	3.938	2	0.140
	Microbrewery	21	83			
	Regional	5	41			
Installed technologies to reduce energy use*	Brewpub	19	47	15.605	2	0.000
	Microbrewery	12	94			
	Regional	2	47			
Use alternative sources of energy*	Brewpub	43	24	12.558	2	0.002
	Microbrewery	61	41			
	Regional	15	31			
Installed water meters to measure and control consumption*	Brewpub	40	26	23.280	2	0.000
	Microbrewery	35	62			
	Regional	8	40			
Reduce water use*	Brewpub	18	47	13.094	2	0.001
	Microbrewery	17	87			
	Regional	1	47			
Recover water	Brewpub	20	45	1.262	2	0.532
	Microbrewery	24	79			
	Regional	11	35			

*Chi-squared significant $p<0.01$

two-thirds of the breweries reported doing this. Spent grains can account for up to 85 % of the brewery's by-product. Regional/regional craft breweries had the lowest percentage, with only 37 % reporting they engaged in or will engage in this practice.

While responses are comparable for collecting and repurposing spent grain, brewpubs and microbreweries are less likely to collected and repurpose their spent yeast. While regional/regional craft breweries are more likely to collect and repurpose spent yeast than their counterparts, the majority (63 %) of them do not or no longer engage in this practice.

Investment in recyclable/reusable packaging and reducing package material scores were comparable. Eighty-six percent of the breweries said they employed or will employ these actions, which bodes well for environmentally sustainable actions. While scores were similar for recovering heat and water, the scores were on the low end with only 20 % and 25 %, recovering heat and water respectively. We expect these scores to increase over time as newer brewing technologies typically afford the ability to recover heat and water during the brewing process.

The installation of technology such as meters to monitor and reduce the consumption of resources (water and energy) also had mixed responses. While most microbreweries (56 %) and regional/regional craft breweries (53 %) installed meters to measure and control energy consumption, the same cannot be said for brewpubs (26 %). The same pattern holds true for installing other technology (not meters) to reduce energy use. While the large majority of microbreweries (88 %) and regional/regional craft breweries (96 %) have undertaken this practice, only 71 % of brewpubs have. Likewise, similar results are apparent for reducing water use, with 84 % of microbreweries and 98 % of regional/regional craft breweries taking such efforts, while only 72 % of brewpubs do the same. These statistically significant differences are most likely due to the dual nature of brewpubs. With both restaurant and brewing operations, it may be more difficult to separate, monitor and reduce the water and energy consumption.

Six environmental sustainable actions were found to have no statistical difference in responses regarding the brewery type. This means all craft brewery types are behaving in a similar manner on these actions (or at least are reporting similar actions). Regarding the collection and repurposing of spent grains, slightly less than two-thirds of the breweries reported doing this. Spent grains can account for up to 85 % of the brewery's by-product. Regional/regional craft breweries had the lowest percentage, with only 37 % reporting they engaged in or will engage in this practice.

While responses are comparable for collecting and repurposing spent grain, brewpubs and microbreweries are less likely to collected and repurpose their spent yeast. While regional/regional craft breweries are more likely to collect and repurpose spent yeast than their counterparts, the majority (63 %) of them do not or no longer engage in this practice.

Investment in recyclable/reusable packaging and reducing package material scores were comparable. Eighty-six percent of the breweries said they employed or will employ these actions, which bodes well for environmentally sustainable actions. While scores were similar for recovering heat and water, the scores were on the low end with only 20 % and 25 %, recovering heat and water respectively. We expect these scores to increase over time as newer brewing technologies typically afford the ability to recover heat and water during the brewing process.

The installation of technology such as meters to monitor and reduce the consumption of resources (water and energy) also had mixed responses. While most microbreweries (56 %) and regional/regional craft breweries (53 %) installed meters to measure and control energy consumption, the same cannot be said for brewpubs (26 %). The same pattern holds true for installing other technology (not meters) to reduce energy use. While the large majority of microbreweries (88 %) and regional/regional craft breweries (96 %) have undertaken this practice, only 71 % of brewpubs have. Likewise, similar results are apparent for reducing water use, with 84 % of microbreweries and 98 % of regional/regional craft breweries taking such efforts, while only 72 % of brewpubs do the same. These statistically significant differences are most likely due to the dual nature of brewpubs. With both restaurant and brewing operations, it may be more difficult to separate, monitor and reduce the water and energy consumption.

Social Sustainability

The final pillar we examined was social sustainability, which is by and large the human component of sustainability. The social sustainability results from the survey are shown in Table 10.6.

Overall the breweries posted comparable scores for most of the social sustainability actions. All three brewery types had high scores for job recognition and for providing a rewarding work environment. This is not surprising as a visit to a local brewery often provides anecdotal evidence of a fulfilling workplace. Interestingly, scores were relatively low for job training and advancement, particularly for brewpubs in which 50 % of the respondents said their brewery does not or no longer does job training and advancement. Again, these answers could also reflect the potential disconnect between those working within the brewery side and those working in the restaurant or bar, or the inherently transient nature of restaurant employment. A further explanation could be the lack of mobility in the workplace, as waiters, for example, typically do not move on to become brewers. Juxtaposed to the low job training opportunities were higher scores on emergency preparedness. Brewpubs, microbreweries and RCBs reported 70 %, 80 % and 85 %, respectively, on providing (or soon providing) such training. Breweries also scored very high on supporting local community groups.

There are three questions in which responses varied by brewery type. First, responses differed considerably with respect to sustainability training for employees. Only 57 % of brewpubs provided or will provide this type of training. This compares to higher scores of 73 % for microbreweries and 90 % for RCBs. Brewpubs may not provide this type of training to employees as not all employees work on the brewing side of operations.

Second, survey answers for employees volunteering for local charity events was significantly lower for brewpubs at 75 %, compared to microbreweries and regional craft breweries, which scored 93 % and 88 % respectively. This might arise from the work cycle of each establishment. With brewpubs, several employees need to be

Table 10.6 Social sustainability Chi squared test results by brewery type

	Type	Used to or does not	Currently does or has plan to	Pearson Chi-square	df	Sig
Job training and advancement opportunities offered	Brewpub	18	47	8.580	2	0.140
	Microbrewery	20	86			
	Regional	3	46			
Recognition for good efforts and potential for increased responsibility	Brewpub	11	54	2.978	2	0.226
	Microbrewery	14	93			
	Regional	3	48			
Rewarding work environment (trustworthy, resources provided etc.)	Brewpub	10	54	4.418	2	0.126
	Microbrewery	10	95			
	Regional	2	47			
Employees are trained on aspects of sustainability*	Brewpub	27	36	14.245	2	0.001
	Microbrewery	29	77			
	Regional	5	43			
Employees volunteer for charity events and social causes*	Brewpub	16	50	10.087	2	0.006
	Microbrewery	7	93			
	Regional	6	41			
Preparedness for emergencies to act promptly and responsibility to minimize impacts to people and the environment	Brewpub	19	46	4.507	2	0.105
	Microbrewery	20	82			
	Regional	6	40			
Locally provided goods and services, as well as living wage jobs etc.	Brewpub	7	59	2.064	2	0.356
	Microbrewery	5	97			
	Regional	3	45			
Support non-profit organizations in the local community	Brewpub	2	65	2.258	2	0.218
	Microbrewery	1	106			
	Regional	0	49			
Organize and fund charitable events*	Brewpub	14	53	13.802	2	0.002
	Microbrewery	10	95			
	Regional	1	48			

*Chi-squared significant $p<0.01$

working during business hours, while with the other two types of breweries, this may not always be the case. Thus employees could have time off to volunteer.

Third, and closely related to the second point, differences were notable for breweries that organized and funded charitable events. While brewpubs had a higher percentage for this social sustainability action (80 %) compared to employee volunteering, they still significantly trailed microbreweries (90 %) and regional craft breweries (98 %). Again, this percentage may be lower because not all brewpub employees work on the brewing side of operations.

In sum, the social sustainability questions sought to shed light on the human dimensions of sustainability. Results indicated similar actions for the three brewery types, with some exceptions. Overall, respondents reported comparable actions regarding job satisfaction, advancement opportunities, and wages. Differences were noted in areas concerning sustainability training and community engagement.

10.5 Conclusion

There are many opportunities for breweries to engage in (more) sustainable practices throughout the brewing process. We have examined how responses from urban craft breweries differ from rural ones (essentially, they don't) and how different types of urban craft breweries (brewpubs, microbreweries and RCBs) have fared regarding sustainability attitudes and actions. The overarching trend was that brewpubs had low scores in most categories (particularly within the environmental and social pillars) when compared to microbreweries or brewpubs. Overall, microbreweries and RCBs scored higher across the board under all three pillars.

From the general sustainability perspective, microbreweries and RCBs were more likely to (1) Adopt technology for reducing water and energy, and (2) Use social media for promoting sustainability in their operations. Further RCBs scored higher with respect to supporting their employees and the local community. The higher scores in these last two aspects reflect the fact that RCBs also had the highest scores for including sustainability in their mission/vision statements.

Under the economic pillar, survey responses did not indicate much difference in attitudes and actions among the brewery types. In fact, the only statistically significant difference that did surface was concerning a systematic review of operations to improve sustainability. Here, regional/regional craft breweries were more apt to conduct such a review compared to microbreweries and brewpubs. Interestingly, both microbreweries and brewpub reported their mission/vision statement contained a commitment to sustainability. Yet, without a review on operations, it could be difficult to ascertain just how well these types of breweries are performing from an economic sustainability perspective.

There were many more differences in attitudes and practice by brewery type under the environmental pillar. While reusing and/or repurposing yeast was largely undertaken by regional/regional craft breweries, less than two-thirds of these breweries actually did so. Microbreweries and RCBs were also more likely to install energy and water meters to monitor and reduce their resource consumption.

Under the social pillar, brewpubs were less likely to offer training to employees on how to be more sustainable compared by microbreweries and RCBs. In addition, the latter two brewery types were more likely to support local charitable events and have employees volunteer at such events.

It should be noted that the purpose of this study was not to determine which type of brewery is most sustainable. Rather, it is to look at sustainability trends for urban craft breweries located throughout the United States. We acknowledge there were some shortcomings in our study, including the relatively low response rate (11 %) and the misclassification of brewery type by respondents. Irregardless, our findings do indicate there is room for improvement for all brewery types regarding sustainability attitudes and actions. We hope future studies build upon our initial findings to determine if sustainability has become further entrenched US craft breweries and if the scale of production influences such trends.

Acknowledgments We would like to thank Ben Brandstetter for sharing his knowledge of beer with us. Thanks to Rob and Sarah for their cross cultural education and understanding of divinely inspired urban breweries. Finally, we are grateful to all the breweries that participated in our survey.

References

Brewers' Association (2015a) Craft beer industry market segments. Accessed 8 June 2015 from https://www.brewersassociation.org/statistics/market-segments/
Brewers' Association (2015b) History of craft brewing. Accessed 8 June 2015 from https://www.brewersassociation.org/brewers-association/history/history-of-craft-brewing/
Brewers' Association (2015c) Number of breweries. Accessed 8 June 2015 from https://www.brewersassociation.org/statistics/number-of-breweries/
Ghiselli RF, La Lopa JM, Bai B (2001) Job satisfaction, life satisfaction, and turnover intent among food-service managers. Cornell Hotel Restaur Admin Q 42(2):28–37
Hoalst-Pullen N, Patterson MW, Mattord RA, Vest MD (2014) Sustainability trends in the regional craft beer industry. In: Patterson M, Hoalst-Pullen N (eds) The geographies of beer. Springer, Dordrecht
Kanagachandran K, Jayaratne R (2006) Utilization potential of brewery waste water sludge as an organic fertilizer. J Inst Brew 112(2):92–96
Lambert EG, Hogan NL, Barton SM (2001) The impact of job satisfaction on turnover intent: a test of a structural measurement model using a national sample of workers. Soc Sci J 38(3):233–250
Mekonnen MM, Hoekstra AY (2010) The green, blue and grey water footprint of crops and derived crop products, Value of water research report series no. 47. UNESCO-IHE, Delft
Olajire AA (2012) The brewing industry and environmental challenges. J Clean Prod 1–21
SABMiller, WWF, GTZ (2010) http://assets.wwf.org.uk/downloads/water__futures.pdf?_ga=1.258501002.1605710182.1433968283. Accessed 1 June 2014
World Commission on Environment and Development (1987) Our common future. Oxford University Press, Oxford, p 27

Chapter 11
Food Hubs: Connecting Farms with Local and Regional Markets

LaDona Knigge, Jacob N. Brimlow, and Sara S. Metcalf

Abstract Food hubs represent an emerging model of food distribution that can connect small and midsize farmers with institutions, schools, and other urban markets by creating new value chains that strengthen the local and regional food system. This study reviews the trends that contribute to the formation of regional food hubs and describes the prospects that they hold for small and midsize farmers. Our discussion is placed in the geographic context of California's North Sacramento Valley using results from a recent survey of farms operating in the region. We conclude that food hubs create a "sweet spot" in matching supply to demand that is instrumental to building a sustainable food system and restoring opportunities for midsize farm operations.

Keywords Food hubs • Local food • Sustainable agriculture • Agriculture of the Middle (AOTM)

11.1 Introduction

There is a growing trend among consumers in the US to buy fresh, locally produced food, much of which is purchased directly from farmers at farmers markets or as a weekly basket of goods through subscriptions to Community Supported Agriculture (CSAs). A 2012 study of 1300 shoppers in the US found that most people prefer to buy local food and are willing to pay more for it (Rushing and Ruehle 2012).

L. Knigge (✉)
Department of Geography and Planning, California State University at Chico,
Chico, CA, USA
e-mail: lknigge@csuchico.edu

J.N. Brimlow
College of Agriculture, California State University at Chico, Chico, CA, USA
e-mail: jbrimlow@csuchico.edu

S.S. Metcalf
Department of Geography, The State University of New York at Buffalo, Buffalo, NY, USA
e-mail: smetcalf@buffalo.edu

J.D. Gatrell et al. (eds.), *Urban Sustainability: Policy and Praxis*,
Geotechnologies and the Environment 14, DOI 10.1007/978-3-319-26218-5_11

Shoppers responding to this survey saw buying local food as helping the local economy (66 %), providing more choices (60 %), and providing healthier food options (45 %) (Rushing and Ruehle 2012). The study also found that shoppers tend to trust locally-owned outlets such as farmers markets, CSAs, locally owned supermarkets, and natural foods stores more than large-scale operations such as national supermarkets and big box retailers (Rushing and Ruehle 2012). This shift toward local food is good news for small and midsize farmers who have been marginalized under the dominance of Big Agriculture.

Trends in the number and size of farms in the US over the years reveal an increasingly bifurcated agricultural and food system, characterized by growth in farming operations at the extremely large and small scales, with declining numbers of medium-sized farms (Low and Vogel 2011). At one end of the spectrum is the expansion of very large, consolidated, industrial agriculture operations that produce most of the food and fiber of the food system. These operations involve production of large quantities of bulk commodities that are marketed through well-established food chains (Kirschenmann et al. 2008). Most agricultural production in the US involves large farming operations under the industrial, market-oriented model of agriculture (Hoppe et al. 2010). This dominant model is characterized by mass-production of economically efficient and federally subsidized commodities such as corn, wheat, and soybeans (Lyson 2004).

At the other end of the spectrum is the proliferation of local, small-scale farming operations that are capable of producing unique, diverse specialty crops and products that are sold to consumers through direct marketing strategies such as sales at farmers markets and through CSAs. Most (91 %) of the farms in the US produced under $250,000 in sales per year as of 2007, comprising approximately one-quarter of the agricultural production in the US (Hoppe et al. 2010). The majority (60 %) of these farms were classified as non-commercial operations with less than $10,000 gross cash farm income (GCFI) per year (Hoppe et al. 2010). Non-commercial, rural-residence family farms signify farming as part of a household's transition to retirement or have operators who are retired or who have other occupations (Hoppe and MacDonald 2013). Off-farm income typically allows these non-commercial farms to operate largely independent of the farm economy (Hoppe et al. 2010, p. iv).

Focusing on commercial farms that could benefit from the presence of a food hub for providing access to local and regional markets, the present study considers three distinct categories of farms by size:

1. *Small farms* or low-sales intermediate farms are commercial operations with gross sales less than $100,000 per year that may be supplemented by off-farm income (Hoppe et al. 2010).
2. *Medium-sized farms* or high-sales intermediate farms have gross farm sales between $100,000 and $250,000 per year and operators who report farming as their major occupation (Hoppe et al. 2010).
3. *Large farms* are defined as those with sales of more than $250,000 per year. These large farms consist of both commercial family farms and nonfamily, industrial farms (U. S. Department of Agriculture National Agricultural Statistics Service 2007).

Most small commercial farms are currently producing commodities such as poultry, beef, hay, grain, or soybeans instead of high-value, specialty-type crops (Hoppe et al. 2010). Nevertheless, small and midsize farms have the potential to produce unique, diverse products and specialty crops according to sustainable agricultural standards that are desired by families seeking healthy, local food options (Kirschenmann et al. 2008, p. 4).

Despite the challenges they face, medium-sized farms have a comparative advantage when it comes to producing unique, differentiated products that consumers desire (Kirschenmann et al. 2008, p. 4). Their size can allow them to be flexible and innovative in the farming methods, marketing strategies, and kinds of crops and agricultural products that are produced. Large, industrial agricultural producers are adept at producing large quantities of commodities of uniform quality. In contrast, very small, non-commercial farmers are limited in the quantity of specialty crops and unique, differentiated products that they can produce. Therefore, it is the small commercial and midsize farming operations that are well positioned to produce greater quantities of high-value, unique, differentiated agricultural products using sustainable agricultural techniques that consumers are demanding. But are the common direct marketing channels such as farmers markets and CSAs the appropriate scale for production capacity of midsize farmers?

Many small and medium-sized farms have been successful at direct to consumer (DTC) marketing channels that include farmers markets, CSA programs, roadside stands, and U-Pick operations. Farms selling through direct marketing channels tend to remain in business longer than those who are not involved in direct markets (Low et al. 2015). While an increase of 180 % in the number of farmers markets from 2006 to 2014 has expanded access of local food for consumers, participation in multiple markets has shown to increase transportation and other costs to farmers and does not necessarily result in higher net profits for small and medium-sized producers (Low et al. 2015, p. 2).

As noted above, medium-size farms have declined amidst the growth trends in farming at the very large and small scales. Medium-size farms face a particular challenge in matching their scale of operation to reachable markets. These medium-size farms tend to produce too much to rely on DTC channels, but not enough to compete financially with corporate, industrial producers in commodity markets characteristic of national and global distribution systems.

The challenge faced by small and medium-size farms in finding markets of the appropriate scale for their operations has been well researched (Lyson et al. 2008; Kirschenmann et al. 2008; Lev and Stevenson 2013). In 2004 the Leopold Center at Iowa State University and the Center for Integrated Agricultural Systems at the University of Wisconsin-Madison, developed a national initiative to promote an "Agriculture Of The Middle" (AOTM). The mission of this initiative is to renew mid-level[1] farms, ranches, and agrifood enterprises that "are declining because they are too small to be served well by commodity markets and too large to be served

[1]The term "mid-level" is used by AOTM to describe a broader category of farms than the term "medium-size" used in the present study.

well by direct markets" (AOTM 2012; Leopold Center of Iowa State University and University of Wisconsin-Madison Center for Integrated Agricultural Systems 2003). The AOTM initiative focuses on business and marketing strategies, changes in public policy, and research and education efforts for farms and ranches based upon size, business organization, and production and marketing strategies. The criteria for AOTM farms include but extend beyond gross annual sales. Defined as farms and ranches typically that are "too big for direct markets but too small for commodity markets," with gross annual sales that may fall between the range of $50K and $500K, depending upon the kinds of crops that are produced, the geography, and the market (AOTM 2012). AOTM operations tend to be family farms that rely upon farming as the main source of household income, with principal operators who claim farming as their main occupation (AOTM 2012).

The AOTM initiative supports the development of "new food business and marketing networks or 'value chains' which link farms/ranches-of-the-middle with food system partners to meet a growing demand for differentiated, high-quality food products" (Leopold Center of Iowa State University and University of Wisconsin-Madison Center for Integrated Agricultural Systems 2003, p. 3). The initiative and others of its kind seek to build scale-dependent value chains between mid-level and small family farming operations with existing food processors, food service distributors, and local or regional grocery stores. The AOTM initiative supports new, emerging models of distribution that share a mutual commitment to "food quality enhancement, environmental and resource stewardship, information transparency, and the equitable sharing of power and economic returns across the value chain" (Leopold Center of Iowa State University and University of Wisconsin-Madison Center for Integrated Agricultural Systems 2003, p. 4).

11.1.1 The Food Hub in Intermediated Market Channels

Food hubs are emerging as important players in intermediated local food market channels, increasing access for small and medium-sized farms by providing services such as aggregation, storage, marketing, distribution, and processing. By participating in intermediated marketing channels, such as sales to restaurants, local grocery stores, schools, hospitals and other institutions, farms utilizing food hubs can increase their local and regional food sales and thereby meet an increasing demand for locally produced food (Low et al. 2015; Low and Vogel 2011).

The USDA defines a regional food hub as "a business or organization that actively manages the aggregation, distribution and marketing of source-identified food products, primarily from local and regional producers to strengthen their ability to satisfy wholesale, retail and institutions demand (Barham et al. 2012, p. 4). Regional food hubs offer a variety of different services beyond aggregation and distribution of locally sourced food that reflect the unique needs and values within particular regions. These services include providing product storage, marketing, transportation and promotional services, as well as training in food safety, production, and post-harvest handling. Services also include assisting with branding and labeling of products, processing including cutting, freezing, and (re)packaging, brokerage services, provisions of facilities such as shared-use kitchens or other production facilities, maintaining facilities for direct retail sales to consumers and handling food donations to local food banks and other emergency food system providers (Fischer et al. 2013; Matson et al. 2013; Barham et al. 2012).

The food hub is a relatively recent addition to the intermediated local food market channel. According to the 2013 National Food Hub Survey, 62 % of the participant food hubs began operations in the last 5 years. The survey identified a variety of different business models and food hub missions. Five categories of food hub business models were identified: nonprofit, for-profit, cooperative, publicly owned and other. The survey found the most financially successful food hubs were for-profit and cooperative in structure, had been in operation for more than 10 years and provided services to a relatively large number of producers (Fischer et al. 2013, p. 48). Additionally, those food hubs in the study that were located in or adjacent to metro areas were more financially independent that those located in nonmetropolitan counties (Fischer et al. 2013, p. 10).

A defining characteristic of food hubs is their ability to provide customers with source-identified products that are locally or regionally produced (Barham et al. 2012, p. 1), which is in contrast to the industrial model of food production based upon the successful development of supply chains that secure large quantities of uniform, standardized agricultural products at competitive prices. The ability of food hubs to communicate information about the origin of the food such as the identity of the farmer, the location of the farm or special practices or circumstances such as sustainable agricultural techniques or the growing of heirloom varieties "carries value that food hubs and producers can potentially use to realize premium prices" (Fischer et al. 2013, p. 7). Depending upon the business model, values beyond premium pricing are part of the mission of many food hubs. Economic, social or environmental values may include providing a fair return for producers, sourcing food from a locally defined area, organic or sustainable growing practices that are environmentally safe, providing access for fresh, healthy food to all community members, supporting food banks and the emergency food system and supporting local economic development. Food hubs are able to extract a higher, more stable price for farmers through transparent value chains that provide customers such as restaurant owners, specialty grocery stores and institutions with the 'authentic farming story' about the products that they purchase and in turn provide to their customers (Lev and Stevenson 2013, p. 2). Studies have shown that consumers want to know more about where and how their food was produced (Brady and O'Brady 2008; Kirschenmann et al. 2008). They want to know that their food is safe and healthy. They are willing to pay more for food that is locally grown with sustainable agricultural practices. They want to know their farmer. Food hubs provide customers with a trusted source of that kind of values-based information.

11.2 Producer Perceptions in California's North Valley

This study draws insights about the prospects for food hubs from a survey of agricultural producers located in Butte, Glenn and Tehama counties in the North Sacramento Valley of California's Great Central Valley. Growers operating in the study area (see Fig. 11.1) were surveyed in 2013 and 2014 to inform the development of a regional food hub in the area. The survey was designed to assess (1) perceptions that agricultural producers have about barriers to selling into local/regional retail and wholesale market channels, (2) producer interest in services that could be provided by a regional food hub, (3) sales channels used by regional farms, and (4) associated farm demographics. The survey instrument was designed by an interdisciplinary team of researchers and based in part on previous studies (ICADS 2009; Peterson et al. 2010).

Survey questions were non-weighted, discrete (yes or no, "select one response," or "select all that apply"), or Likert Scale ratings. The survey questions used in this study, a subset of the full survey, are listed in the Appendix. The survey was pilot tested by a convenience sample of area farmers to assess reliability and validity. Pilot surveys were given in person and over the telephone, and feedback from the pilot surveys were incorporated into the final instrument. A random sample of growers was drawn from publicly available farmer lists provided by the county departments of agriculture in Butte, Glenn, and Tehama counties. Surveys were administered over the telephone using a standard oral script, and contacts were screened to ensure that they produced food for sale and not only for personal consumption. Of the 200 surveys completed, the present study is based upon the 175 surveys from farms located within Butte, Glenn, and Tehama counties only.

The map in Fig. 11.1 highlights the locations of the three counties – Butte, Glenn, and Tehama – that comprise the study area. The total population is indicated for each county, pointing to the significance of Butte county with 220,000 residents. The smaller populations of Glenn and Tehama counties reflect the importance of farming in the region.

The pie charts shown in Fig. 11.1 show the relative prevalence of small farms (gross sales <$100k/year), medium-sized farms (gross sales $100–$250k/year), and large farms (gross sales >$250k/year) at the county scale, along with number of participant farms in each category. These categories are consistent with the USDA farm typology in use when the survey was designed (Hoppe and MacDonald 2013). The USDA typology has recently been updated to reflect the national trends described above, now allowing for further distinctions among very large-scale farms, and accounting for increases in the consumer price index (Hoppe and MacDonald 2013).

These results are also reported in Table 11.1, which shows the relative prevalence of farms answering the producer perception survey by size category. In response to the question, "How large is your farm? (Gross sales per year)," participant farms indicated whether their size was "Small (less than $100,000 per year)," "Medium ($100,000–$250,000 per year)," or "Large (greater than $250,000 per year)."

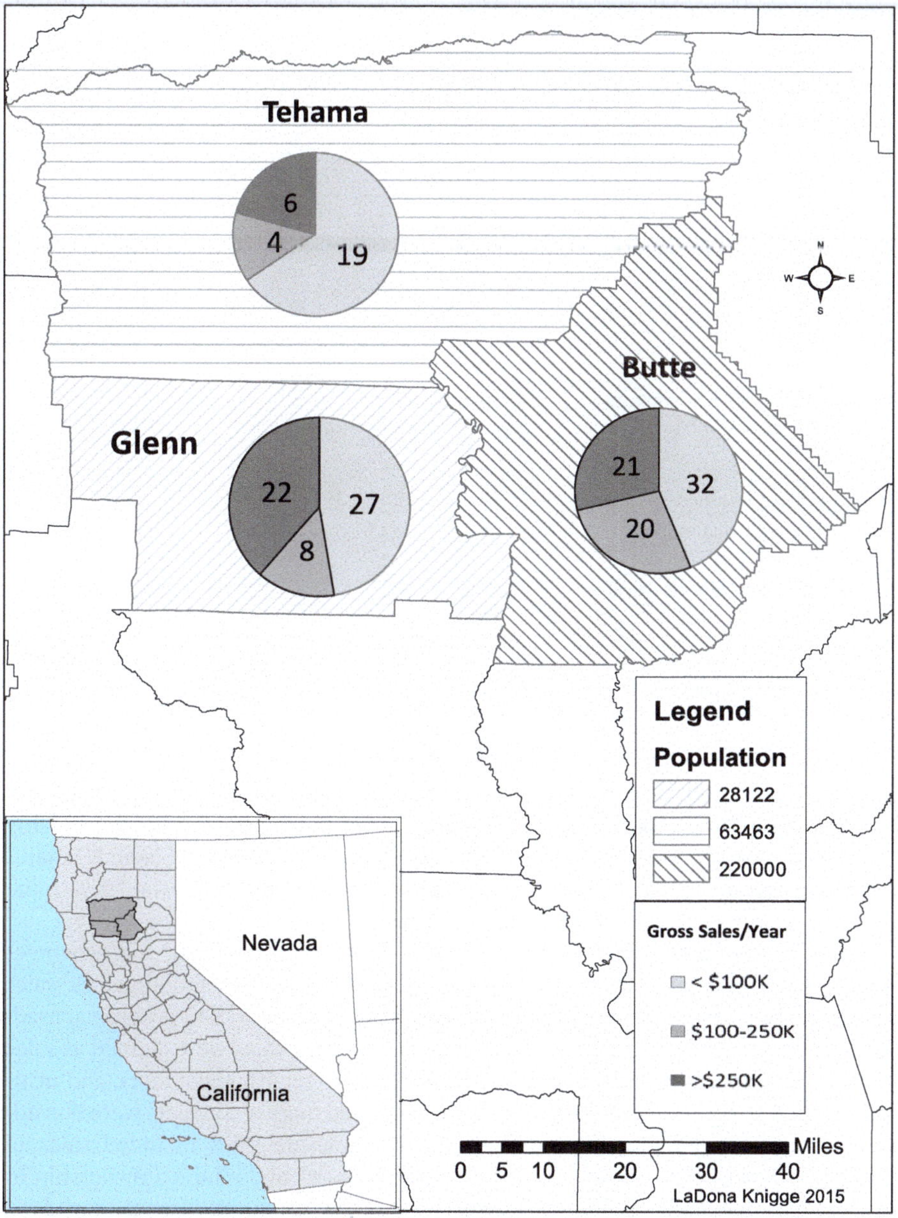

Fig. 11.1 County population and annual gross sales of participant farms

Table 11.1 shows that 16 farms (9 % of the total) declined to state the scale of their operation and are thus excluded from the pie charts in Fig. 11.1.

The results in Table 11.1 indicate that small farms reporting annual sales of less than $100,000 were the most frequent (45 %) overall, while large farms with annual

Table 11.1 Size of participant farms in annual gross sales

	Butte (n=82)	Glenn (n=62)	Tehama (n=31)	Total (n=175)
Small	32	27	19	78
(<$100k/year)	(39 %)	(44 %)	(61 %)	(45 %)
Medium	20	8	4	32
($100k–$250k/year)	(24 %)	(13 %)	(13 %)	(18 %)
Large	21	22	6	49
(>$250k/year)	(26 %)	(35 %)	(19 %)	(28 %)
Declined to state	9	5	2	16
	(11 %)	(8 %)	(6 %)	(9 %)

Table 11.2 Market channels used by participant farms

	Butte (n=82)	Glenn (n=62)	Tehama (n=31)	Total (n=175)
Sell direct to consumers (local)	22 (27 %)	12 (19 %)	6 (19 %)	40 (23 %)
Sell in intermediated markets (local)	13 (16 %)	9 (15 %)	1 (3 %)	23 (13 %)
Direct and intermediated (local)	12 (15 %)	3 (5 %)	1 (3 %)	16 (9 %)
Sell wholesale	63 (77 %)	56 (90 %)	25 (81 %)	144 (82 %)
Wholesale and other (local)	11 (13 %)	14 (23 %)	4 (13 %)	29 (17 %)

Responses and percentages (shares) are reported for producers answering in the affirmative to each question

sales over $250,000 were the second most frequent in the sample at 28 %. Medium-size farms represented the smallest share at 18 %. Farm size varied significantly across counties. For example, 61 % of Tehama County respondents selected "small," compared to only 39 % in Butte County, and Butte County had over 10 % more farms selecting "medium" than Glenn and Tehama.

Farmers were asked whether they sell into direct, intermediated, and/or wholesale markets channels. Direct to consumer (DTC) sales were defined as sales through famers' markets, community supported agriculture (CSA) programs, roadside stands, and U-pick operations. Intermediated market sales were defined as sales to grocery or specialty retail stores, retail food coops (retail), restaurants, and institutions such as schools. Both direct and intermediated market channels were considered "local" for the purposes of this study. Wholesale sales included sales to growers' cooperatives, wholesale buyer/broker/packers, and retained ownership or commission merchants. Table 11.2 reports the results from producer responses to this part of the survey.

The surveyed farms indicated sales that were concentrated in the wholesale market channel, but significant numbers of farms also sold in direct and intermediated channels. A majority (82 %) of producers sold through wholesale channels, 23 % sold into direct channels, and 13 % of farms sold into intermediated channels. Butte County had the highest share of farms selling into direct (27 %) and intermediated (16 %) market channels. Butte County has a regional comprehensive university

(California State University, Chico), the most active farmers markets in the region, the largest population, and highest population density at 130 people per square mile as of the 2010 U.S. Census.

To the extent that diversified marketing strategies represent local and regional food sales, results of the survey do not necessarily support a common perception that local and regional food systems develop more quickly or completely in areas of higher population density. Glenn County has the region's smallest population and a population density of only 21 people per square mile as of the 2010 U.S. Census, and had the highest percentage of farms reporting wholesale sales. However, Glenn County also had the highest share of farms reporting sales in both wholesale and local (direct and/or intermediated) channels (23 %). Overall, 20 % of surveyed farms sold in multiple market channels, and 17 % reported diversified sales into both wholesale and local channels.

Sales in direct and intermediated market channels can bring higher prices than wholesale, so there are potential benefits for large farms that can be derived from selling a portion of their production into these channels. However, a lack of economies of scale in local small-scale distribution, along with high search and contracting costs, often make smaller volume local sales prohibitively expensive. Further, the challenges of coordination and cost of allocating production to two different market channels or to multiple direct markets can be significant (same as found in Low et al. 2015).

Participant farms were asked to indicate whether they agreed with each of a set of statements described a major or minor barrier to selling products or expanding their businesses (see Appendix). As shown in Table 11.3 below, 27 % of surveyed farmers reported a barrier in being too large for small buyers, likely reflecting the challenges just described. As expected, the share of farms in a county reporting being too large for small buyers reflects size distribution: Glenn County has the largest share of farms reporting that the quantities they produce are too large for small buyers (32 %) and also has the largest share of large farms (35 %; see Table 11.1).

In contrast, Table 11.3 indicates that 18 % of surveyed farms reported the challenge of the quantities they produce being too small for large buyers. Many small farms have more production than they can sell in direct markets, but have difficulty

Table 11.3 Scale and volume barriers among participant farms

	Butte (n=82)	Glenn (n=62)	Tehama (n=31)	Total (n=175)
Production is too small for large buyers/distributors	15 (18 %)	9 (15 %)	7 (23 %)	31 (18 %)
Production is too large for small businesses	23 (28 %)	20 (32 %)	4 (13 %)	47 (27 %)
Production is both too small for large buyers and too large for small buyers	4 (5 %)	2 (3 %)	2 (6 %)	8 (5 %)

meeting requirements set by wholesale buyers such as volume, food safety certification, and packaging. Again, survey results reflect farm size distribution in each county: Tehama County has the largest share of farms reporting being too small for large markets (23 %) and also has the largest share of small farms in the sample (61 %; see Table 11.1).

The producer perceptions survey asked farmers if they would be interested in paying for food hub services designed to help facilitate increased local/regional sales, including cold/dry storage, packaging and light processing/washing, aggregation, marketing/brokering, and food safety (see Appendix). A majority (60 %, 105 farms) of the 175 farms surveyed indicated an interest in one or more food hub services. The highest share of farms interested in food hub services was in Glenn County (66 %, 41 farms), followed by farms in Butte County (61 %, 50 farms), with the least interest reported in Tehama County (45 %, 14 farms). Despite this variation, the results of the survey indicate considerable interest in the benefits that food hubs can provide to both small and large farms by providing aggregation and coordination services to connect small farms to larger volume markets, and by reducing search and contracting costs to facilitate sales from large farms into higher price point local markets.

11.2.1 The North Valley Food Hub

In late 2014, an emerging food hub, the North Valley Food Hub (NVFH),[2] began facilitating transactions between regional agricultural producers and intermediated market buyers such as restaurants and institutions. The NVFH was born out of 5 years of targeted local food systems program development in the tri-county region of Butte, Glenn, and Tehama counties. Figure 11.2 illustrates the location of the NVFH relative to road infrastructure spanning this tri-county region of Northern California.

A feasibility study conducted early in the NVFH's development showed that the business would not be financially viable if significant fixed costs were incurred to support hub infrastructure (Northern California Regional Land Trust). The NVFH began transacting in 2014 without warehouse or transportation infrastructure, accomplishing distribution using a "pop-up" site (Hacking 2015). The photograph in Fig. 11.3 shows the mobile unit that constitutes this pop-up site at the California State University, Chico University Farm located south of the city.

The mobile cold-storage unit depicted in Fig. 11.3 allows NVFH staff to store product on a short-term basis as needed. Growers deliver purchased products to the site in the morning (see Fig. 11.4), NVFH staff consolidate orders and prepare invoices, and buyers pick up or arrange for third party delivery later the same day. Without warehousing and transportation infrastructure, the primary value to

[2]The NVFH website is http://northvalleyfoodhub.com/ and in turn links to an interactive GIS map for the foodshed of NVFH growers in the region: http://geochico.csuchico.edu/bfblnv/

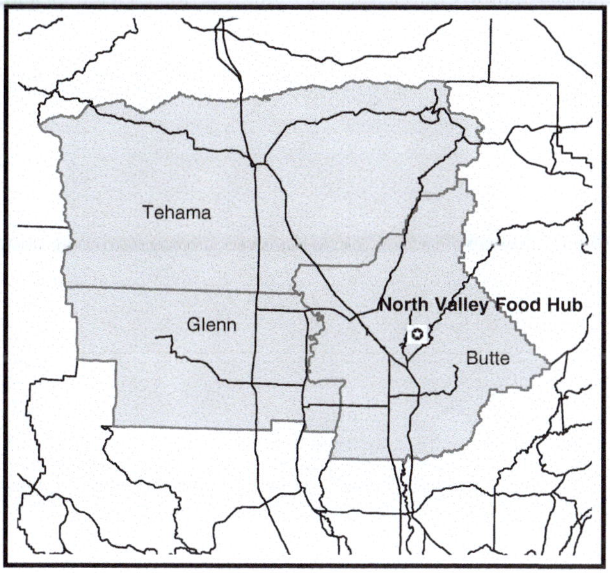

Fig. 11.2 The location of the NVFH in central Butte County

Fig. 11.3 The mobile unit of the NVFH pop-up site at the University Farm

Fig. 11.4 An NVFH drop-off transaction

customers provided by the NVFH derives from reduced search/coordination time, streamlined purchasing/invoicing, and product aggregation.

11.3 Discussion

Food hubs offer opportunities for producers, particularly small and medium-size farmers, to expand their markets, stabilize prices and increase their profits. While direct to consumer market channels offer higher prices for farmers, they do not support sufficient sales volume for many midsize farmers. Because of the size of their operations, they are too large for direct markets, but also too small to compete successfully against large farms in the commodity markets. Food hubs provide aggregation and distribution services as well as a variety of additional services to participating producers and are becoming an integral part of local and regional food systems across the United States.

Studies have identified several challenges for the long-term sustainability and financial success of food hubs. The 2013 Food Hub Survey found balancing supply and demand and managing growth to be the biggest challenges for the food hubs in the survey. Part of the challenge is associated with seasonal operation as well as the need for more efficient operations or effective management. Food hubs that operated seasonally reported difficulties in finding reliable seasonal or part-time staff (Fischer et al. 2013). Both the 2013 Food Hub Survey and a study of non-profit food hubs in Vermont found reliance on volunteer labor to be problematic (LeBlanc et al. 2014). The Vermont study also found that food hubs in rural areas were less likely to extract a premium price for local food than food hubs in urban areas. Access to capital, finding appropriate technology and meeting regulatory requirements were also cited as challenges by respondents to the 2013 Food Hub Survey (Fischer et al. 2013, pp. 40–41). On the positive side, once assured more stable markets through participation in the food hub, many producers were reported to take on practices to extend their growing season, diversified their product offerings, increased the amount of acreage in production and many adopted more sustainable agricultural practices such as integrated pest management (Fischer et al. 2013, p. 17).

11.4 Conclusion

Food hubs are emerging players in intermediated marketing channels for local and regional food, and support the diversification, resilience, and long-term sustainability of local food systems. By offering distribution, aggregation, and other services, and maintaining information about food source identification and growing practices, food hubs provide value to farmers, food retailers, and consumers. Farmers utilizing food hubs to sell regionally often benefit from higher and more stable prices than those available in wholesale markets. Enhanced economic security from participation in local/regional markets can enable farmers to increase the number of acres farmed and the amount of food produced. Food hubs thereby represent a potentially sustainable strategy for small and medium-size farmers to diversify their sales in additional markets that were previously unattainable. As active and dynamic entities, food hubs can effectively expand the market for local food into restaurants, small grocery stores, schools, and other institutions that typically reply upon wholesale distributors.

Acknowledgments The producer survey described in this chapter was supported by a California State University Agricultural Research Institute (ARI) grant and was conducted in conjunction with the CSU, Chico, Center for Healthy Communities.

11.5 Appendix: Survey Questions

How large is your operation? (Gross sales per year)			
Small (less than $100,000 per year)	❑	Large (greater than $250,000 per year)	❑
Medium ($100,000 - $250,000 per year)	❑	Decline to state	❑

Answer the following with regard to your product sales in 2012-13. Please ✓ select Yes or No.		
Did you sell to any of the following direct consumer outlets?	**Yes**	**No**
Farmers' market	❑	❑
CSA (Community Supported Agriculture)	❑	❑
Roadside stand/farm store	❑	❑
U-Pick sale	❑	❑
Other direct sales:	❑	❑
Did you sell to any of the following groups or third parties?	**Yes**	**No**
Growers' cooperative	❑	❑
Wholesale buyer/broker/packer	❑	❑
Retained ownership or commission merchant	❑	❑

Did you sell to any of the following retail outlets?	**Yes**	**No**
Grocery store	❑	❑
Food cooperative	❑	❑
Restaurant	❑	❑
Institution (such as schools and hospitals)	❑	❑
Other:	❑	❑

Food hubs can help local growers combine and sell products to buyers by improving supply, storage, packaging and delivery (distribution). Fees are charged for services like those listed below. If a food hub was available to you, which of the following services would you consider paying to use? (You may ✓ select up to four)			
Cold storage	❑	Combining products for big volume buyers	❑
Dry Storage	❑	Brokering	❑
Packaging	❑	Marketing	❑
Light processing (trimming and/or peeling)	❑	Food Safety Services (consulting /education)	❑
Washing	❑	Other:	❑

Please ✓ indicate if any of the following statements describe a major or minor barrier to selling products or expanding your business. If the statement does not describe a barrier for your business, please ✓ select "Not a Barrier." ***[Tell farmer he/she will also be asked for the most important barrier]***			
	Major Barrier	**Minor Barrier**	**Not a Barrier**
Quantities you produce are too small for large buyers/distributors	❑	❑	❑
Quantities you produce are too large for small businesses like restaurants	❑	❑	❑

Note: this is a subset of the barriers presented in the original question as asked in the full survey.

References

AOTM (Agriculture of the Middle) (2012) Characterizing Ag of the middle and values-based food supply chains. http://www.agofthemiddle.org/archives/2012/01/characterizing.html. Accessed 19 Jan 2015

Barham J, Tropp D, Enterline K, Farbman J, Fisk J, Kiraly S (2012) Regional food Hub resource guide. United States Department of Agriculture (USDA) Agricultural Marketing Service, Washington, DC

Brady E, O'Brady C (2008) Consumer considerations and the agriculture in the middle. In: Lyson TA, Stevenson G, Welsh R (eds) Food and the mid-level farm: renewing an agriculture of the middle. The MIT Press, Cambridge, pp 103–117

Fischer M, Hamm M, Pirog R, Fisk J, Farbman J, Kiraly S (2013) Findings of the 2013 national food hub survey. Michigan State University Center for Regional Food Systems & The Wallace Center at Winrock International, Lansing. www.ngfn.org/resources/ngfndatabase/knowledge/2013%20Food%20Hub%20Survey%20Report.pdf

Hacking H (2015) Chico food hub is up and running, on a small scale. Chico Enterp Rec http://www.chicoer.com/general-news/20150211/chico-food-hub-is-up-and-running-on-a-small-scale. Accessed 13 Feb 2015

Hoppe RA, MacDonald JM (2013) Updating the ERS farm typology, EIB-110, U.S. Department of Agriculture, Economic Research Service. http://www.ams.usda.gov/sites/default/files/media/Regional%20food%20hub%20Resources%20Guide.pdf. Accessed 1 Feb 2015

Hoppe RA, MacDonald JM, Korb P (2010) Small farms in the United States: persistence under pressure. U.S. Department of Agriculture, Economic Research Service, Washington, DC. http://www.ams.usda.gov/sites/default/files/media/Regional%20food%20hub%20Resources%20Guide.pdf

ICADS (Intervale Center Agricultural Development Services) (2009) Expanding local food production, storage and marketing capacity in Vermont: results from the 2007 farm producer survey. Intervale Center Agricultural Development Services, Burlington. www.intervale.org/wp-content/uploads/IC-2009-Expanding_Local-Food-Production.pdf

Kirschenmann F, Stevenson G, Buttel F, Lyson TA, Duffy M (2008) Why worry about the agriculture of the middle? In: Lyson TA, Stevenson G, Welsh R (eds) Food and the mid-level farm: renewing an agriculture of the middle. MIT Press, Cambridge, pp 3–22

LeBlanc JR, Conner D, McRae G, Darby H (2014) Building resilience in nonprofit food hubs. J Agric Food Syst Commun Dev. Advance online publication. http://dx.doi.org/10.5304/jafscd.2014.043.005

Leopold Center of Iowa State University and University of Wisconsin-Madison Center for Integrated Agricultural Systems (2003) Renewing an agriculture-of-the-middle: a three phase national initiative, www.intervale.org/wp-content/uploads/IC-2009-Expanding_Local-Food-Production.pdfBurlington

Lev L, Steenson G (2013) Values-based food supply chains: an introduction to nine case studies. The National Initiative on Agriculture of the Middle, Ames

Low SA, Vogel S (2011) Direct and intermediated marketing of local foods in the United States. ERR-128, U.S. Department of Agriculture, Economic Research Service, Washington, DC, November 2011

Low SA, Adalja A, Beaulieu E, Key N, Martinez S, Melton A, Perez A, Ralston K, Stewart H, Suttles S, Vogel S, Jablonski B (2015) Trends in U.S. local and regional food systems. APO68, U.S. Department of Agriculture, Economic Research Service, Washington, DC, January 2015

Lyson TA (2004) Civic agriculture: reconnecting farm, food and community. Tufts University Press, Medford

Lyson TA, Stevenson GW, Welsh R (eds) (2008) Food and the mid-level farm: renewing an agriculture of the middle. MIT Press, Cambridge

Matson J. Sullins M, Cook C (2013) The role of food hubs in local food marketing. U.S. Department of Agriculture (USDA) rural development service report 73

Northern California Regional Land Trust (n.d.) Feasibility study: food hub – creating local food systems. Morrison & Co. http://morrisonco.net/case-studies/northern-california-regional-land-trust/. Accessed 6 May 2015

Peterson HH, Selfa T, Janke R (2010) Barriers and opportunities for sustainable food systems in northeastern Kansas. Sustainability 2:232–251. doi:10.3390/su2010232

Rushing J, Ruehle J (2012) Buying into the local food movement. A.T. Kearney, New York. https://www.atkearney.com/documents/10192/709903/Buying+into+the+Local+Food+Movement.pdf/68091049-b5c2-4d2a-a770-ee5b703da8fd

U. S. Department of Agriculture National Agricultural Statistics Service (2007) 2007 census of agriculture: small farms. U.S. Department of Agriculture, Washington, DC

Chapter 12
Is Urban Sustainability Possible in Post-Katrina New Orleans?

Wendy A. Lascell

Abstract Hurricane Katrina struck the city of New Orleans on August 29th, 2005 forever changing the *Big Easy* and its hard hit neighborhoods (Fig. 12.1). The storm surge breached the inadequately engineered levees in several locations flooding 80 % of the city under 1 to over 10 ft of water (NOAA 2005; Waple A, Hurricane Katrina. NOAA's National Climatic Data Center, Asheville, 2005; Baumann PR, Nellis MD, Mitteager WA, Geocarto Int 21(2):75–80, 2006; Plyer A, Facts for features: Katrina impact. Report published at www.datacenterresearch.org. The data center: independent analysis for informed decisions in Southeast Louisiana, 2014). It became the largest residential disaster in United States history (Plyer A, Facts for features: Katrina impact. Report published at www.datacenterresearch.org. Independent analysis for informed decisions in Southeast Louisiana, The Data Center, 2014), testing New Orleans' resilience. In fact, most discussions about the future of the city have been about resiliency (Campanella TJ, J Am Plan Assoc 72(2): 141–146, 2006; Gotham and Greenberg 2014). While the ability to recover is crucial for the future of the city, the idea of sustainability has to be considered. Will New Orleans be rebuilt with sustainability ideals of social justice and equity or will racial discrimination and poverty prevail?

Historical conditions of the socio-political ecology are riddled with examples of racial segregation (Germany KB, J Am Hist 94(3):743–751, 2007) leading to extreme vulnerability of black neighborhoods (Bond Graham 2007) showcased by the events following Hurricane Katrina. In the initial few years following the flood event, many predicted New Orleans would be "whitened" (Fussell E, Sastry N, VanLandingham M, Popul Environ 31(1):20–42,2010) as displaced black residents did not have the means to return and seemingly would never be able to return (Logan JR, The impact of Katrina: race and class in storm-damaged neighborhoods. Unpublished manuscript. Spatial structures in the social sciences initiative, Brown University, 2006; Horne J, Five myths about Hurricane Katrina, The Washington Post, 31 August, 2012). However, although black residents have returned at much slower rates (Fussell E, Sastry N, VanLandingham M, Popul Environ 31(1):20–42,2010), the feared "whitening" of New Orleans has not come to fruition, preserving much of the cultural characteristics of the city.

W.A. Lascell (✉)
Geography Department, SUNY Oneonta, Oneonta, NY, USA
e-mail: Wendy.Lascell@oneonta.edu

J.D. Gatrell et al. (eds.), *Urban Sustainability: Policy and Praxis*, Geotechnologies and the Environment 14, DOI 10.1007/978-3-319-26218-5_12

Keywords Sustainability • New Orleans • Social justice • Hurricane Katrina • Urban geography

12.1 Background

The city of New Orleans consists of 74 neighborhoods, segregated largely by race and secondarily by income. Early in its development, the nickname "Crescent City" emerged illustrating the sliver of development of the high grounds along the

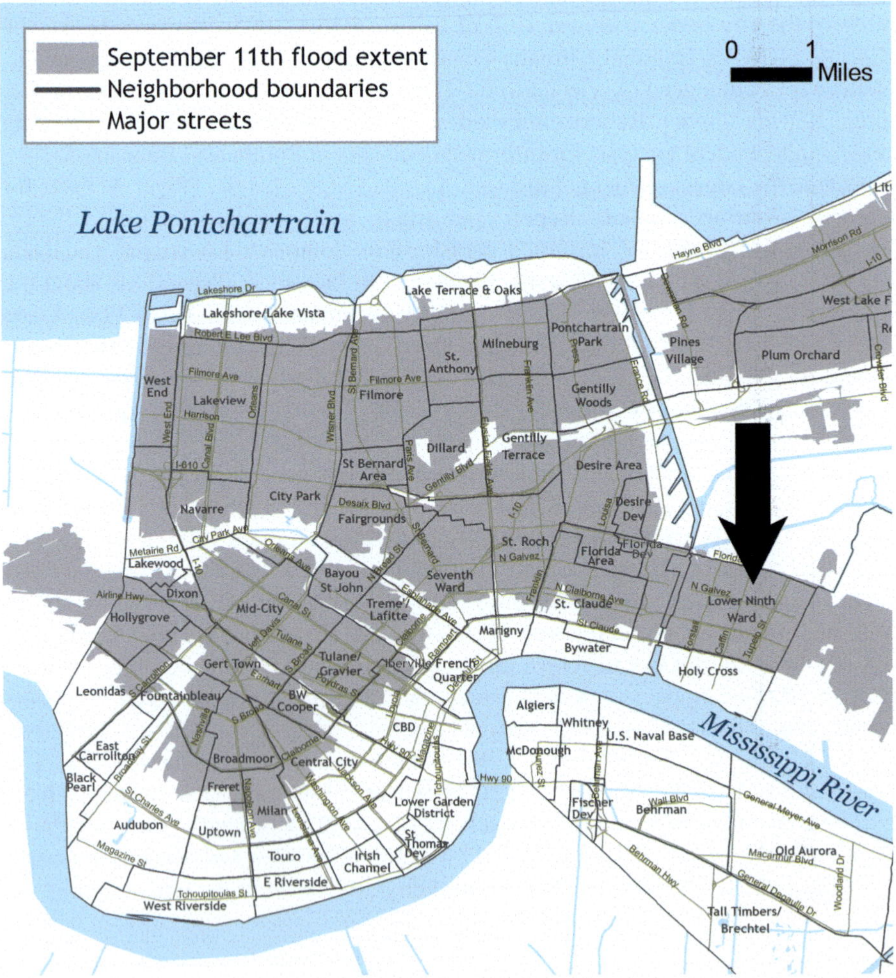

Fig. 12.1 Map of New Orleans, showing the location of the study area. The Lower 9th Ward neighborhood was flooded following Hurricane Katrina (Adapted from The Data Center 2005)

Mississippi River. Over a long period of time (1730s–1870s and 1920s–1960s), the lower elevations were filled-in and developed and the engineering infrastructure of levees and pumping stations helped keep low-lying areas dry (Sills et al. 2008). Prior to Katrina, Hurricane Betsy caused the largest flooding event in 1965 but it did not cause widespread damage (Germany 2007) creating a sense of security among New Orleans residents. Extreme vulnerability to hurricanes was a serious issue due to the physical conditions, tenuous levee infrastructure, and social-economic characteristics of the city residents (Cutter and Emrich 2006). Although faced with many hazards such as land subsidence, coastal wetland erosion, and potential levee failure (National Resources Defense Council 2011) the city's residents lived in these low-lying neighborhoods seemingly blind to its vulnerability.

Arguably the most socially and physically vulnerable neighborhood in the city of New Orleans is the Lower 9th Ward (Fig. 12.2). Historically an impoverished and segregated neighborhood, it lies several feet below sea level and is located adjacent to the Industrial Canal. According to the 2000 U.S. Census, the Lower 9th Ward housed 14,008 people in 4,820 households, 98.3 % of which were Black or African American. Household incomes of less than $20,000 annually were reported by 50.5 % of the households in the Lower 9th Ward. Residents of this neighborhood are less likely to have access to vehicles than other New Orleans residents. All of these help create an extreme vulnerability to catastrophic events.

Also known as the "Lower 9" the neighborhood contained mostly small, one-story, wooden shotgun style homes (Fig. 12.3) sitting noticeably below levee walls along the Industrial Canal, which was providing only an illusion of safety. Residents relied on the engineering of this floodwall system maintained by the Army Corps of Engineers. Failing to properly consider the soil types and understand the complexities of the landscape, levees were not built to sufficient standards to withstand Katrina's strong Category 3 Hurricane storm surge (Horne 2012; Seed et al. 2005; Sills et al. 2008). Two of the levee breaches allowed the storm surge to forcefully stream into the Lower 9th Ward (Fig. 12.4) causing widespread destruction (Fig. 12.5). If the levees hadn't failed, the discussion about post-Katrina New Orleans would be quite different.

12.2 Sustainability

Sustainability is a complex concept, which since its inception has typically referred to considerations of the natural environment and development. However, sustainability within an urban context must include discourse of socio-economic and political ecologies. As the consideration of "social sustainability" of urban places increases, it remains a new concept with limited literature and lack of agreement on a definition or specific goals (Dempsey et al. 2009). In this paper, social justice and equity will be discussed in the context of New Orleans and specifically the Lower 9th Ward to explore the possibilities of socially sustainable communities.

Fig. 12.2 QuickBird satellite image, 2003. Resolution ≈60 cm. Pre-Katrina landscape of the Lower 9th Ward neighborhood (DigitalGlobe 2003)

Plenty of academics, scientists, and political figures have hesitated to encourage rebuilding in New Orleans based on physical geographies such as the elevation of the city below sea level and the likelihood for increased frequencies and intensities of storms in the future. The risk of vulnerability is high and many say that certain neighborhoods such as the Lower 9th Ward should not be rebuilt. Controversial as it is, from a strictly environmental perspective, rebuilding in these vulnerable neighborhoods might be equivalent to flushing money down the proverbial toilet. It seems unlikely, however, that any portion of these vulnerable neighborhoods will be converted to parks and wetlands. Residents want to rebuild. They want to come "home" to their beloved city even if it is a struggle to recreate their pre-Katrina existence. But should those rebuilding simply recreate what previously existed? Great opportunities exist for cities forced to rebuild after widespread disasters. Why not create a *new* New Orleans instead of settling on simply renewing what previously existed? The possibilities are exciting. The question remains – can New Orleans do it? Can they reach sustainability goals of social justice and equity despite the environmental challenges?

Fig. 12.3 Typical small, one-story, wooden structures in New Orleans neighborhoods. Floodwater depth exceeded the height of the one-story homes in the Lower 9th Ward (Photo taken by Wendy A Lascell, November 19, 2007)

Most rebuilding has focused on resilience instead of sustainability. Homes have been gutted and repaired, demolished, and rebuilt while many lots remain empty after widespread government demolishment of homes. Empty lots await residents who may not be able to rebuild from scratch because of economic restraints. Lack of reconstruction in neighborhoods such as the Lower 9th Ward is cited as an example of an environmental justice issue (Maantay and Maroko 2009).

Fortunately, a variety of rebuilding projects throughout the city have been underway, relying mostly on aid from nonprofits and volunteers. The lucky ones received Road Home monies to rebuild from the controversial and largely criticized government program (Green and Olshansky 2012). Habitat for Humanity's Musicians' Village in the Upper 9th Ward consists of houses raised up above Katrina flood levels meant to ensure the livelihood of the music culture (Fig. 12.6). Area churches have housed volunteers from around the country to help with gutting and rebuilding all over the city. Common Ground Collective's relief program based in the Lower 9th Ward has been helping local residents with legal aid, medical care, gutting and rebuilding, wetland and soil restoration, and general community support since immediately after the storm. Tulane University has been utilizing their architecture graduate program to help design rebuilding projects. And arguably the most famous rebuilding project located in the Lower 9th Ward is the Make It Right foundation founded by actor Brad Pitt. All of these rebuilding projects contribute to resiliency and some contribute to the sustainability of New Orleans.

Fig. 12.4 The levee wall failed and water forcefully streamed into the Lower 9th Ward. A barge is seen that floated through the levee breach, destroying homes in its path. The current location of Common Ground Collective is seen near the barge. The Mississippi River is in the background (Photo taken by Augustino 2005)

12.3 Social Justice and Equity Concerns

Black communities in New Orleans have historically been discriminated against and have suffered atrocities such as police brutality and institutionalized racism. There are countless examples illustrating the conflict along racial and socio-economic lines in New Orleans. In the 1960s New Orleans was experiencing extreme segregation and inequality, remarkably low education rates, severe impoverishment, and high rates of violence including a rate of murder almost twice the national average (Germany 2007). The birth of the New Orleans chapter of the Black Panther Party (1970) was partly due to racial conflict persisting in the 1960s. It was co-founded by Malik Rahim who also later co-founded the grassroots relief organization Common Ground Collective based in the Lower 9th Ward to assist residents with a variety of needs following Hurricane Katrina.

When the federal government finally forced desegregation of schools in a highly resistant 1960s New Orleans, public schools were defunded through continuous tax cuts. "Resegregation" followed as private schools opened in the city for white children resulting in over 80 % black children in substandard public schools by the 1970s (Piliawsky 1979). Continuous white flight to the suburbs left behind concentrations of poor black populations in neighborhoods across the city such as the

Fig. 12.5 WorldView-1 satellite image, 2013. Resolution ≈50 cm. Much of the Lower 9th Ward remains desolate. The small area (near the *bottom left* of image) with new construction is the Make It Right neighborhood

Lower 9th Ward (Germany 2007; Bond Graham 2007) served by low-quality and low-performing schools taken over by the state after 2005 (Johnson-Burel et al. 2014). In post-Katrina New Orleans an attempt via formation of Charter Schools is being made in order to establish integration and racial equity in the educational system (Johnson-Burel et al. 2014; Horne 2011). Perforated with both criticism and support from residents and politicians, success of this new system is yet to be determined.

While there are plenty of truths to the discriminatory history of the black population in New Orleans, myths or urban legends persist and color the perceptions of those living in poverty (Horne 2012). Most of the Lower 9 residents encountered by the author over a period of 2006–2012 stated with conviction that the government

Fig. 12.6 A newly constructed home in the Habitat for Humanity's Muscians' Village in the Upper 9th Ward, New Orleans (Photo taken by Wendy A Lascell. May 24, 2007)

intentionally bombed the levees, in order to save the French Quarter and spare disaster for the "rich people." This urban legend is based on a somewhat altered historical event – the bombing of levees along the Mississippi River during the Great Flood of 1927. The current residents say, "they have done it before" and truly believe that the government bombed the levees in 1927 to eradicate the poor black population in the Lower 9th Ward. The truth is the levees were bombed farther downstream and both black and white residents were evacuated by the government prior to the breach. However, there was conflict created as those evacuated were promised compensation that was never received or was substantially inadequate (Groen and Polivka 2008). Influenced by many stories passed through generations current residents of the Lower 9th Ward feel neglected and abandoned by their own government. Compounded by former President G.W. Bush and his administration's apparent lack of concern about the wellbeing of the people of New Orleans, the attitudes of distrust have become even more deeply entrenched.

The Urban Land Institute (2005) issued a plan shortly after the storm's impacts were assessed suggesting the city shrink in size to be sustainable in the long-term, by banning the redevelopment of hard-hit neighborhoods and concentrating the population in functioning areas. While from an urban planning perspective this is the most reasonable and intelligent response to a catastrophe of this size, residents saw it as restrictive and unfair (DePillis 2013). This plan could have contributed both positively and negatively to the sustainability of the city as it embodies contradictory ideals. It would make the city more efficient and prevent the rebuilding and maintenance of infrastructure in desolate areas while shutting down historically poor black neighborhoods with a strong sense of community.

Many black residents believe the people in power locally want to rid the city of the poor, black population but maintain its cultural identity (Bond Graham 2007) for the tourism economy. Initially the Bring New Orleans Back Commission (BNOB) formed by Mayor Ray Nagin endorsed the Urban Land Institute's report (Meitrodt and Donze 2005) and shared its own proposal that residents would have 1 year to rebuild and if no substantial progress was evident, entire neighborhoods would be demolished and redeveloped by the city for both urban development and green space restoration. Given that BNOB was largely made up of white business people with a large number of real estate investors, it was clear to the community that developers were foaming at the mouth to get control of the Lower 9th Ward considering its proximity to the French Quarter and Industrial Canal. While this may not be true and the intentions of the BNOB were possibly genuine, it was seen as a social injustice not as smart planning and the prevention of urban blight.

A legal team, partly organized by Common Ground Collective, fought these deadlines in court delaying the mass demolition until the summer of 2008. Mayor Nagin was forced to back down from the idea of seizure and redevelopment of the demolished space and the BNOB issued its final report in January of 2008. Social justice took a big hit as homes in the Lower 9th Ward were eventually demolished in careless fashion, in many cases without confirmation that homes were actually abandoned. Residents responded by using spray paint to convey messages to the bulldozing teams, communicating that homes were actually being worked on and provided phone numbers to call for confirmation (Fig. 12.7) since their efforts to work with City Hall were largely unsuccessful mostly due to severe communication issues within city government and its contractors. It is hard to know for sure if certain neighborhoods were targeted for this FEMA-funded demolition (Verderber 2009) for reasons other than abandonment and health hazards (from mold growth). Some homeowners returned home to an empty lot not knowing their house was even being considered for demolition (Krupa and Lolis 2007).

It was thought that with the displacement of so many black residents and widespread demolition of homes, New Orleans would be "whitened" and some of the core culture lost. Certainly it was more difficult for low-income black residents to return home so they returned at slower rates than those of higher socioeconomic status (Fussell et al. 2010). Recent data shows that the percent black has not decreased by as much as predicted. Pre-Katrina black population was at about 67 % and post-Katrina black population is estimated at about 60 % (Horne 2012). However, if individual neighborhoods are examined it is clear that many black residents have not returned – at least not to the Lower 9th Ward as much of that neighborhood still remains desolate. September 2008 data of residences receiving mail in the Lower 9th Ward show that less than 5 % of residents had returned (Baumann and Mitteager 2009) while other parts of the city had rebounded faster (Fussell et al. 2010). While the black population slowly has been increasing, the 2010 U.S. Census records just 2842 people in the Lower 9th Ward compared to 14,008 prior to Katrina.

Fig. 12.7 Homeowners responded to mass demolitions by spray-painting messages on their homes in efforts to deter bulldozers (Photo taken by Wendy A Lascell, November 19, 2007)

What is the future of the Lower 9th Ward neighborhood? It is difficult to predict but certainly the future will be influenced by two key nonprofits located in the neighborhood: Common Ground Collective and the Make It Right Foundation. Both have adopted a grassroots approach to aiding the recovery of the Lower 9th Ward. Their contributions are substantial and will most likely continue into the future.

12.4 Common Ground Collective

Immediately after Hurricane Katrina and the consequent levee failures caused havoc in New Orleans, Malik Rahim a local activist and resident of New Orleans co-founded Common Ground Collective with two other activists. They opened up a home once lived in by Rahim's mother in the Lower 9th Ward and began helping local residents with recovery and social justice issues. Free medical and legal clinics were established, a food pantry was opened, and they began helping people sort through the messes they encountered at their homes. A volunteer movement was born, bringing in countless numbers of mostly college-aged volunteers from all over the country. More commonly known as Common Ground Relief (CGR) they arranged volunteer housing at a Catholic school in the Upper 9th Ward. The pastor shared space with the organization and in turn CGR helped with getting the church reopened and fixing the school. The author led volunteer trips taken by the SUNY Oneonta student-groups to volunteer with CGR joining many others from all over the country in helping residents recover.

At first, recovery from chaos was the priority. Residents struggled with returning to a city without open schools and hospitals or functioning infrastructure. Important documents such as deeds were lost forever as no copies were to be found anywhere including City Hall. CGR has gutted over 3,000 homes (Common Ground Relief 2014) and helped countless residents with gaining access to their properties. Assisted by CGR, one resident prepared for doomsday, focused on building hundreds of boats out of recycled foam materials so he could "save (his) people" when the next storm hit. As time passed, ideals such as solidarity, community, and social justice were identified and set as goals of CGR. Smaller subsets of CGR were formed such as the Media Collective and the Soil Remediation team. The author was part of the Media Collective in 2006 and helped establish linkages with media for communication purposes. Common Ground Collective continues to operate in the Lower 9th Ward and no end is in sight. Steady streams of volunteers and substantial donations keep the work successful. Working towards social justice and equity among the black and poor populations, CGR is contributing to the future social sustainability of New Orleans.

12.5 Make It Right Foundation

The headquarters of the Make It Right Foundation is also located in the Lower 9th Ward, a stone's throw from the Common Ground Relief headquarters. While Common Ground Relief has worked in many neighborhoods across the city, Make It Right focuses its energies on the Lower 9th Ward only. In this neighborhood, home-ownership was nearly 60 % (Campanella 2006; Plyer 2014) prior to Hurricane Katrina tearing it to shreds. Identified as a crucial rebuilding site for historically underserved populations, Make It Right was founded after Brad Pitt visited the Lower 9 in 2007 and the first six homes were built during the summer of 2008. The organization has currently built approximately 100 homes and they are committed to building a total of 150 (Make It Right 2014; Verderber 2009). The homes constructed with local laborers are built with green initiatives in mind, contributing to sustainability of the area through community consultations and innovative materials. Homes are raised up and stand out among the older and traditional architecture of New Orleans.

Make It Right has raised approximately $45 million (DePillis 2013) mostly for building the Lower 9 homes and their strategy is currently a source of controversy. At first, it was clear that these homes would be habited by former Lower 9 residents displaced by the storm with no means of returning otherwise. Recently, they have opened the process up to non-Lower 9 residents (Make It Right 2014) because of a lack of interested residents (DePillis 2013).

But how does Make It Right contribute to sustainability goals? Positive contributions include green living and available housing for displaced impoverished residents of the Lower 9th Ward – at least those that want to return. Social justice and equity are served for the black population that call the Lower 9 "home." Perhaps the money could have been spread out more to provide additional housing and it could have been located in other neighborhoods less vulnerable to future storms. The Make It Right foundation gets plenty of criticism about these particular points. It is the author's belief that while it would be more efficient urban planning to largely abandon the Lower 9 and concentrate efforts on a "shrunken" city, it would not be a *just* decision. A compromise between the two options seems to be the best decision: provide infrastructure to the Lower 9th Ward for a renewed neighborhood but only allow a smaller sized neighborhood to be built. Restrict construction to the southeast corner of the neighborhood to prevent a spatial pattern of scattered low-density homes costing more for the city to serve. Unfortunately, that ship has sailed. There didn't seem much room for compromise and services have been restored to the entire low-density area. If an error had to be made, at least it was in favor of historically underserved populations – that is, for now.

12.6 Urban Planning

Overall, the city government has backed down and not made any grand planning decisions for the city. Rebuilding and renovation efforts have been left to individual decisions of organizations and residents ostensibly steeped in corruption with little transparency (Verderber 2009). Lack of informed planning is leading to controversy. A few of the planning decisions made by the city on a small-scale have been obstructions to social justice and equity. For example, housing "projects" have been torn down in the name of Katrina damage, even though they never flooded (Hammer 2008). Even the United Nations spoke out on the decision to demolish the HUD-sponsored housing for its violation of basic human rights and racial discrimination (United Nations 2008). Although many residents surveyed did not want to return to the housing projects slated for demolition (Hammer 2008), the city was not including sufficient replacement housing in their plans (United Nations 2008).

The city did publish GreeNOLA (2008), a sustainability plan for the city focused on topics such as waste management, flood risk reduction, clean energies, and green building. Social justice and equity were not considered, leaving large gaps in this master plan document. In 2013 the New Orleans Business Alliance, under the guise of Mayor Mitch Landrieu, published ProsperityNOLA. This is a plan for creating jobs and economic growth with discussion of sustainability and "nurturing its human, economic and natural resources" (NOLABA 2013). Again, here is another example of an initiative that provides a great opportunity to discuss the social justice and equity for the people of New Orleans. Instead it focuses on economic growth, which tends to help business owners and not necessarily the worker bees. The jury is still out, it is doubtful this new plan will lead to any advances in social sustainability.

12.7 Summary

The inadequate response to this disaster reveals social inequities and vulnerability of the impoverished residents (Cutter 2006) in neighborhoods such as the Lower 9th Ward. Hurricane Katrina disproportionately impacted poor black communities in New Orleans. Lack of concern by policy-makers and even the former President sends a clear message to New Orleans black population that they do not matter. During the State of the Union Address in January of 2007, George W. Bush failed to mention Hurricane Katrina or New Orleans. The only remark to the Gulf region at all was in reference to the Middle East (Germany 2007). If New Orleans as a city is largely ignored, how is the Lower 9 supposed to recover with social sustainability?

Large expanses of the neighborhood remain uninhabited with overgrown lots and few services. The efforts of Common Ground Collective and especially the Make It Right Foundation grow the community with new housing meant for displaced residents. Many of those in the Make It Right homes are elderly as the younger generation opts for hipster neighborhoods instead (DePillis 2013). Considering 75 % of the pre-Katrina population has not returned to this neighborhood, can it regain its character and become a functional community once again? Is social sustainability possible? Hopefully the neighborhood will redeem itself without restoring high crime and drug rates. Perhaps after the Make It Right Foundation has completed its project and all homes are filled, the community will be a shrunken and older version of itself. It is unlikely that social sustainability will be achieved without unrelenting grassroots efforts aimed at the best interests of black populations essential to the cultural core of New Orleans. Hurricane Katrina has commemorated its 10-year anniversary, but it will take at least another decade for the Lower 9th Ward to settle into its new urban character.

Note Wendy A Lascell formerly published under the name Wendy A Mitteager; she has spent the last decade studying New Orleans spending substantial time volunteering for Common Ground Collective.

References

Augustino J (2005) Aerial views of damage caused by Hurricane Katrina, the day after the hurricane hit. FEMA. 30 August 2005. https://www.fema.gov/media-library/assets/images/45236

Baumann PR, Mitteager WA (2009) Analyzing Katrina's impact on New Orleans' Lower 9th Ward with high-resolution satellite imagery. SUNY Oneonta 9th Annual Faculty Research Show, 12 February 2009

Baumann PR, Nellis MD, Mitteager WA (2006) Katrina: the "Big One" arrived. Geocarto Int 21(2):75–80
Bond Graham D (2007) The New Orleans that race built: racism, disaster, and urban spatial relationships. Souls 9(1):4–18, http://cretscmhd.psych.ucla.edu/nola/volunteer/EmpiricalStudies/The%20new%20orleans%20that%20race%20built%20-%20racism,%20disaster,%20and%20urban%20spatial%20relationships.pdf
Campanella TJ (2006) Urban resilience and the recovery of New Orleans. J Am Plan Assoc 72(2):141–146
Common Ground Relief (2014) Rebuild New Orleans., From, http://www.commongroundrelief.org/about-us
Cutter SL (ed) (2006) Hazards, vulnerability and environmental justice. Earthscan, London
Cutter SL, Emrich CR (2006) Moral hazard, social catastrophe: the changing face of vulnerability along the Hurricane coasts. Ann Acad Polit Soc Sci 604:102–112
Dempsey N, Bramley G, Power S, Brown C (2009) The social dimension of sustainable development: Defining urban social sustainability. Sustainable Development 19:289–300
DePillis L (2013) If you rebuild it, they might not come: Brad Pitt's beautiful houses are a drag on New Orleans, New Republic, 13 March 2013
DigitalGlobe (2003) QuickBird satellite image. Resolution ≈60 cm
Fussell E, Sastry N, VanLandingham M (2010) Race, socioeconomic status, and return migration to New Orleans after Hurricane Katrina. Popul Environ 31(1):20–42
Germany KB (2007) The politics of poverty and history: racial inequality and the long prel ude to Katrina. J Am Hist 94(3):743–751
Green TF, Olshansky RB (2012) Rebuilding housing in New Orleans: the road home program after the Hurricane Katrina disaster. Hous Policy Debate 22(1):75–99
Groen, JA, Polivka AE (2008) Hurricane Katrina evacuees: who they are, where they are, and how they are faring. Monthly Labor Review (March 2008). http://www.bls.gov/opub/mlr/2008/
Hammer D (2008) Survey: most don't want to return to public housing. The Times-Picayune, 6 March 2008. http://www.nola.com/news/index.ssf/2008/03/survey_most_dont_want_to_retur.html
Horne J (2011) New schools in New Orleans: school reform both exhilarated and imperiled by success. Educ Next 11(2):15–24. www.educationnext.org
Horne J (2012) Five myths about Hurricane Katrina. The Washington Post. 31 August 2012
Johnson-Burel D, Drame E, Frattura E (2014) Participatory research in support of quality public education in New Orleans. Educ Action Res 22(3):288–305
Krupa M, Lolis EE (2007) Protecting their own. The Times-Picayune, 10 August 2007. http://blog.nola.com/updates/2007/08/protecting_their_own.html
Logan JR (2006) The impact of Katrina: race and class in storm-damaged neighborhoods. Unpublished manuscript. Spatial structures in the social sciences initiative, Brown University
Louisiana Disaster Recovery Foundation (2008) GreeNOLA: a strategy for a sustainable New Orleans: a resource for neighborhoods & residents to engage in the master planning process and ensure sustainable redevelopment. Master Plan, City of New Orleans
Maantay J, Maroko A (2009) Mapping urban risk: flood hazards, race, & environmental justice in New York. Appl Geogr 29:111–124
Meitrodt J, Donze F (2005) Plan shrinks city footprint. The Times-Picayune, 14 December 2005
National Resources Defense Council (2011) New Orleans, Louisiana: identifying and becoming more resilient to impacts of climate change. From https://www.nrdc.org/water/files/ClimateWaterFS_NewOrleansLA.pdf
NOAA (2005) Hurricane Katrina., From, http://www.ncdc.noaa.gov/extremeevents/specialreports/Hurricane-Katrina.pdf
New Orleans Business Alliance (NOLABA) (2013) ProsperityNOLA: a plan to drive economic growth for 2018, executive summary. New Orleans. www.nolaba.org
Gotham KF, Greenberg M (2014) Crisis cities: Disaster redevolopment in New York and New Orleans. Oxford University Press, New York

Piliawsky M (1979) Public education: best hope for the "New South". J South Reg Coun 1(8):5–7
Plyer A (2014) Facts for features: Katrina impact. Report published at www.datacenterresearch.org. Independent analysis for informed decisions in Southeast Louisiana, The Data Center
Seed RB, Nicholson PG, Dalrymple RA, Battjes JA, Bea RG, Boutwell GP, Bray JD, Collins BD, Harder LF, Headland JR, Inamine MS, Kayen RE, Kuhr RA, Pestana JM, Siva-Tulla F, Storesund R, Tanaka S, Warman J, Wolff TF, Wooten RL, Zimmie TF (2005) Preliminary report on the performance of the New Orleans levee systems in Hurricane Katrina on August 29, 2005. Report No. UCB/CITRIS, 17 November 2005. University of California at Berkeley/ Center for Information Technology Research in the Interest of Society
Sills GL, Vroman ND, Wahl RE, Schwanz NT (2008) Overview of New Orleans Levee failures: lessons learned and their impact on National Levee design and assessment. J Geotech Geoenviron 134(5):556–565
The Data Center (2005) New Orleans area – maps. Map created with data from FEMA, The City of New Orleans Planning Commission, and Census TIGER. Independent analysis for informed decisions in Southeast Louisiana, The Data Center. http://www.datacenterresearch.org/maps/
United Nations (2008) UN experts call for protection of housing rights of Hurricane Katrina victims. UN News Centre, 28 February 2008. www.un.org/apps/news/story.asp?NewsID=25782
Urban Land Institute (2005) New Orleans, Louisiana: A strategy for rebuilding., From http://uli.org/wpcontent/uploads/2012/11/NewOrleans-LA-05-v5.pdf
Verderber S (2009) The unbuilding of historic neighbourhoods in post-Katrina New Orleans. J Urban Des 14(3):257–277
Waple A (2005) Hurricane Katrina. NOAA's National Climatic Data Center, Asheville

Chapter 13
Mapping Environmental Justice: A Framework for Understanding Sustainability at the Neighborhood Scale in Indianapolis

Trevor K. Fuller

Abstract Sustainability has at its core the three strands of environment, economy, and social equity. Environmental justice, as it has progressed and evolved since the early 1980s, has come to be envisioned as encompassing these same components. A community that is low-income, minority-majority population, and houses a greater proportion of a city's environmental hazards (and/or less of amenities like parks as well), can be seen as a community faced with challenges of building equity across environmental, social, and economic lines. Using an environmental justice framework I examine how effective policies and practices are, or have been in producing 'sustainable development' or a 'sustainable community'. I assess the City of Indianapolis' efforts to transform a historically disadvantaged environmental justice community into a 'sustainable community' and examine environmental justice within the city as a means by which to evaluate 'sustainable development'. Additionally, Geo-technologies (i.e., GIS) were used to provide a visualization of the potential environmental risks (and amenities) within the city's neighborhoods. The risk of a redevelopment approach rooted in economic conceptions of sustainability as opposed to social and environmental conceptions will likely produce a narrowly 'sustainable', but broadly gentrified community.

Keywords Environmental justice • GIS • Sustainable community • Indianapolis

Over the past two decades 'Sustainability', and its myriad associated terms, has increasingly become a prominent part of 'government-speak' throughout local, state, and federal levels of government in the U.S. The term 'sustainable development' was coined in the now infamous "Our Common Future", a 1987 United Nations report prepared by the Brundtland Commission. At its very core are three goals: social equality, economic vitality, and environmental quality. While

T.K. Fuller (✉)
Department of Geography and Environmental Sustainability,
State University of New York at Oneonta, Oneonta, NY, USA
e-mail: fullertk@oneonta.edu

J.D. Gatrell et al. (eds.), *Urban Sustainability: Policy and Praxis*,
Geotechnologies and the Environment 14, DOI 10.1007/978-3-319-26218-5_13

developed within a global framework, in practice, sustainable development must be envisaged and operationalized at the 'local' scale. Many municipal governments across the U.S. have recognized the need for such an approach. However, moving from policy to practice has proven difficult. While many policies have been created concerning 'sustainable communities' or 'smart growth', it continues to be difficult for governments to realize all three components of sustainability as originally conceived. What we find are instances in which cities are more easily able to produce change in the 'environmental' portion of the sustainability triad as compared to the economic and social portions. While this is problematic when in pursuit of a broader notion of sustainability, it does offer hope for environmental justice communities.

Environmental justice is an inherently spatial project. It originated from the discovery that economically and socially disadvantaged communities tend to host a disproportionate share of environmental hazards relative to other communities. In the United States, where the environmental justice movement began, this spatial project manifested itself in communities in which low-income and/or minority populations are unevenly burdened by the presence of environmental hazards, hazards often unseen in higher-income Caucasian communities. The hazards disproportionately present in low-income/minority communities can range from large and potentially toxic facilities such as Superfund sites to more localized activities like illegal dumping sites. Beyond the distributive strand of environmental justice is participatory or procedural justice. This developed later in the progression of environmental justice literature, which long focused on distributive questions only. Procedural justice is the notion that low-income and/or minority-majority communities dealing with environmental hazards should be provided a venue in which their voice, their concerns, can be heard and included in the decision-making process.

In recent years there has been a surge of interest and scholarship regarding the 'nexus between environmental justice and sustainability' (Agyeman 2005, 2008). This stems from the connections between two previously disparate frameworks. Sustainability has at its core the three strands of environment, economy, and social equity. Environmental justice, as it has progressed and evolved since the early 1980s, has come to be envisioned as encompassing these same components. A community that is low-income, minority-majority population, and houses a greater proportion of a city's environmental hazards (and/or less of amenities like parks as well), can be seen as a community faced with challenges of building equity across environmental, social, and economic lines. Through the environmental justice framework we can begin to examine how effective policies and practices are or have been in producing 'sustainable development' or a 'sustainable community'. Here we approach the City of Indianapolis' efforts to transform a historically disadvantaged environmental justice community into a 'sustainable community' and examine environmental justice within the city as a means by which to evaluate 'sustainable development'. Additionally, we use Geo-technologies (i.e., GIS) to provide a visualization of the potential environmental risks (and amenities) within the city's neighborhoods.

13.1 Study Area: Background and Opportunity

The broader study area is the city of Indianapolis, Marion County, Indiana, a conservative Midwestern city of approximately 820,000 people (US Census 2010). In terms of socioeconomic characteristics, 61.8 % of the city's population is white and 27.5 % African-American. Additionally, the median household income between 2008 and 2012 was $42,144. Nearly 20 % of the city's population lives below the poverty level (US Census 2010). Indianapolis has experienced several periods of population and economic growth since its establishment in 1820 (Fig. 13.1). However, the second half of the twentieth century saw a consistent decline in the population of its city center (Hulse and Zeigler 1991). The population of Center Township during the 1970s declined by approximately 24 % (id).

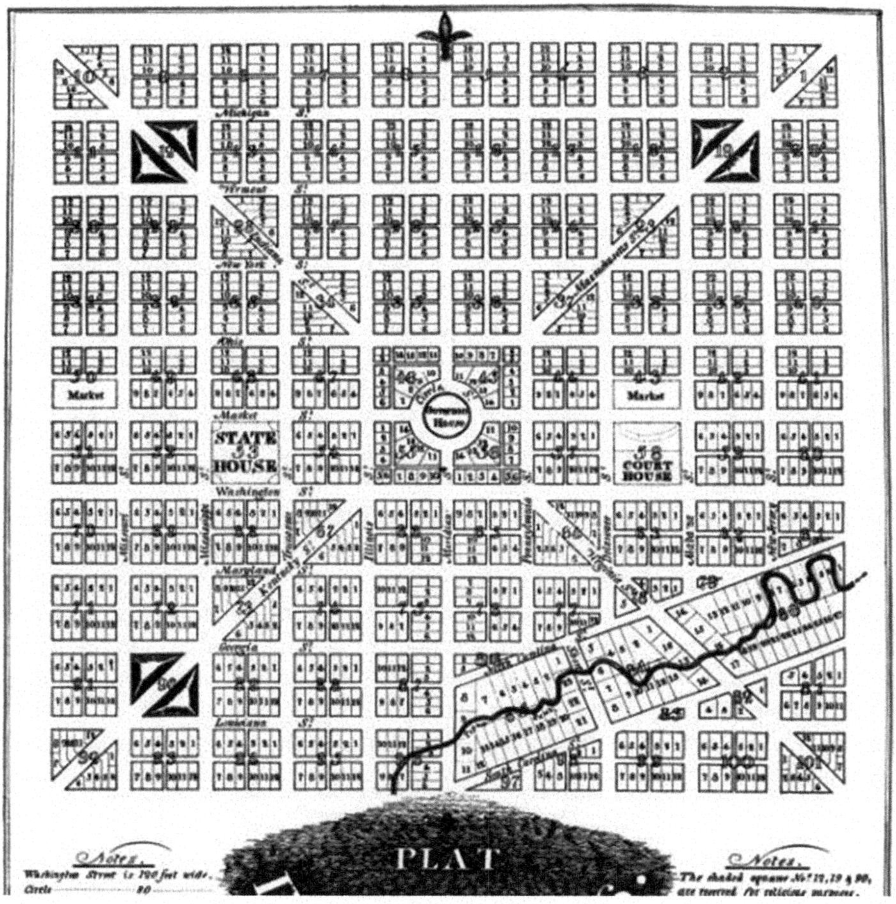

Fig. 13.1 Plat of the first 'mile square' layout of Indianapolis in 1820 (Indiana Historical Society)

Martindale-Brightwood, a neighborhood that has been the focus of previous research (Fuller 2014), as well as the center of the City of Indianapolis' sustainability measures, comprises a relatively small area on the near northeast side of Indianapolis (Martindale-Brightwood Community History Profile 2014). In a fashion similar to that of the city center, Martindale-Brightwood saw the white population move to the distant suburbs, leaving behind housing ultimately occupied by low-income African-American residents (Hulse and Zeigler 1991). The neighborhood suffered another blow in 1976 with the development of two interstates (I-70 and I-65) directly through it, thereby displacing several residents, cutting the community in half, and offering a new environmental hazard with its associated noise, air, and soil pollution. For much of its recent history Martindale-Brightwood residents have been faced with unemployment, housing needs, and industrial pollution. In particular, Martindale-Brightwood has served as an unwilling host of a slew of manufacturing and industrial processes, along with their associated environmental hazards. The presence of so many hazardous facilities in a relatively small community has left a legacy of environmental contamination that persists to today.

In 2004, an environmental site assessment in the neighborhood revealed historical practices which may have produced a potential for site contamination. After further investigations were conducted, high levels of lead (Pb) in the soil surface were discovered at the assessed site and across over 200 residential properties. The high levels of contamination propelled some in the community to form an activist organization (Martindale-Brightwood Environmental Justice Collaborative) with the intent to learn more about the status of their neighborhood, as well as to push for additional state and federal cleanup efforts. The Martindale-Brightwood Environmental Justice Collaborative (MBEJC) pushed city, state, and federal officials to 'cleanup' the neighborhood. Remediation of the contaminated soil eventually occurred with some residents left wondering why their property had not been remediated in contrast to their neighbor's yard. The MBEJC eventually partnered with a local nonprofit organization in order to pursue (and eventually secure) US EPA funding. This funding was initially used to foster the efforts of the MBEJC in gathering community input and concerns regarding the environmental conditions in the area. In particular, the residents and the MBEJC were focused on pushing the EPA to perform additional remediation in the area. However, the non-profit organization was essentially being paid by MBEJC via the US EPA grant funding. This brought to light a conflict of interest in using EPA funding to ultimately challenge the EPA remediation effort (which has redirected some of the initial MBEJC agenda). Additionally, the City's influence is playing out currently as the 'revised' agenda of the MBEJC (sans push for additional lead remediation) offered support for the City's goal of redeveloping Martindale-Brightwood as a 'sustainable community'.

13.2 Pursuing Sustainability: Environmental Justice as an Analytical Framework

Environmental justice has evolved over its relatively young history from a purely distributive notion of justice wherein low-income and/or minority communities are burdened by environmental hazards more than other parts of a given area (SAR International, Inc. (1987); UCC 1987; Bullard 1990; Lee 1992; Mohai and Bryant 1992; Anderton 1994; Pulido 2000; Chambers 2007; Checker 2008) to a broadly defined framework encompassing distribution, political representation/participation, and quality of life in general. In particular, its proponents continue to argue in favor of a broader framework, one which casts the 'environment' as not simply environmental hazards, but rather, "where we live, where we work, and where play"(Alston 1991). This newly cast 'environment' allows an expansion of environmental justice to include the social and economic inequities also present in communities housing a disproportionate share of a city's environmental hazards. It forces a focus on not just toxic release sites and the like, but those release sites in concert with issues of racism, unemployment, affordable housing, and general disinvestment (Warner 2002). When the scope of environmental justice moves environmental risks to include social, economic, and racial equity, then the bridging of environmental justice with sustainability becomes even more apparent. In redeveloping or reimagining previously disinvested neighborhoods decision-makers must incorporate environmental, social, economic, and racial equity (id) into their future plans.

Despite sharing several commonalities in terms of buzzwords, the two themes of sustainability and environmental justice often end up straying from one another when implemented. While environmental justice advocates have long pursued a sustainability which incorporates the broader notion of the 'environment' offered by an environmental justice framework, one is hard-pressed to find sustainability policies/initiatives which specifically mention, let alone incorporate, issues such as equity, justice, and inclusion (Lubitow and Miller 2013). While environmental justice has expanded its reach to incorporate social and economic inequities, sustainability continues to be focused on "infrastructural interventions aimed at enhancing environmental outcomes." (Id, 121). What is promoted in the sustainability policies is the much less politicized forms of 'sustainable development', such as redeveloping a brownfield site (regardless of reuse), encouraging economic development in the form of redeveloped sites which simply continue the cycle of not employing area residents. Redeveloping a brownfield site as a 'mini-park' or green space is important as well, though it is simply another example of sustainability without the politics. In this way, sustainability and its measures such as brownfields, area 'apolitical', something assumed to be beneficial for all. With the focus on 'economic redevelopment', and a very narrow form of it at that, the more deeply politicized issues of race and inequality are avoided by city officials (Lubitow and Miller 2013). Rather, particularly visible projects seen as unquestionably characteristics of a 'sustainable community', are easily promoted in favor of the much more complex characteristics of a socially and economically just community.

Environmental justice provides an appropriate framework for assessing sustainability efforts because it has long argued for environmental justice to include "the range of social and economic priorities that continue to impair the health and well-being of low-income and minority communities" (Lubitow and Miller 2013, 126). Agyeman et al. (2002) introduced the notion of 'just sustainability' to ultimately push for this integration of environmental justice with sustainability. The goal of such a notion is a form of sustainability which incorporates "wider questions of social needs and welfare, and economic opportunity" (Agyeman et al. 2002, 78).

The City of Indianapolis was selected in (2008) to be one of five cities awarded funding for a pilot project to enhance a city's ability to create a 'sustainable community'. Funding came from the "Partnership for Sustainable Communities", a federal government partnership between three agencies; US EPA, US Department of Housing and Urban Development (HUD), and the US Department of Transportation (DOT) (USEPA et al. 2010). Nowhere in the "Partnership for Sustainable Communities" literature will one encounter discussions of race, equity, or justice. The pilot project in Indianapolis was geared towards brownfields redevelopment ("Partnership for Sustainable Communities EPA Brownfields Pilot" 2011). The study area, Martindale-Brightwood, was selected as the focus area for this pilot project for good reason as it has historically held the greatest number of brownfields per capita compared to the rest of the city's neighborhoods. Martindale-Brightwood is a large part of the overall "Smart Growth Redevelopment District" created by the City of Indianapolis in 2008. The area itself is rife with unemployment, environmental hazards (primarily brownfields), infrastructural issues, and a property vacancy rate of approximately 33 % (compare to 9 % of the city as a whole). The city has produced a 'magnet school' (The Project School), a few brownfields redevelopment success stories, including one that became an urban garden, and a litany of press touting its sustainability efforts. The irony is that in previous research in this neighborhood a resident speaking in regards to the widespread lead contamination (and partial remediation) said "we can't even have a garden to grow things because of lead" (Fuller 2014).

Martindale-Brightwood houses more brownfields than any other community in the city (Indianapolis Department of Metropolitan Development 2009). As of 2010, Martindale-Brightwood contained approximately 60 brownfields. The City of Indianapolis considers there to be an additional 84 sites of concern that may be classified as brownfields in the near future. Brownfields are properties which are abandoned, unused, or underused and have yet to be redeveloped due to potential or existing environmental issues. Since the late 1990s, brownfields have become a popular redevelopment 'tool' of local, state, and federal levels of government because of the potential benefits associated with redevelopment of such sites – to provide jobs, reduce blight, address existing environmental hazards, and return sites to the tax rolls. In one sense, the abandoned or unused nature of a brownfield site means it will not be subject to any standard schedule of regulatory oversight and monitoring. In fact, the primary way such a site will become known to regulatory parties is if a potential buyer of the site applies for assistance from city, state, or federal brownfields programs. An additional way for such a site to be 'discovered' is if the city decides to conduct an inventory of brownfields. Of course, even in that case, the environmental issues or hazards associated with the site will only be addressed if a buyer or redeveloper is found. This abundance of brownfields stems from decades of disinvestment in the area on the part of the city. Area residents must wait for investment to become possibly fruitful before capital returns and brownfields are redeveloped. Only when sites are financially and legally attractive do we then discover the environmental hazards associated with a site.

Brownfields are an economic development tool first and foremost. Environmental issues, if any, are addressed in order to alleviate liability concerns of prospective developers. Given the inequitable number of brownfields in Martindale-Brightwood as compared to the rest of Indianapolis, and the significant association between race and brownfields, communities such as Martindale-Brightwood are at the whim of government and/or developers in terms of discovering and remediating brownfields environmental contamination. The City of Indianapolis has long recognized the economic benefits of brownfields redevelopment. It is in this way that communities such as Martindale-Brightwood experience injustice both in terms of the initial inequitable distribution of pollution sources (particularly brownfields) and the redevelopment-dependent process of discovering the inequitable contamination produced decades ago.

The City of Indianapolis has long focused on promoting growth within its downtown core at the cost of neglecting its outlying neighborhoods, an approach that brings with it gentrification (Wilson 1991). Conservative pro-growth coalitions – property owners, elites and corporate interests, along with pro-development city agencies – drive much of the location and form of redevelopment in Indianapolis (id). As a result, Indianapolis 'development' acquires form via promotion of corporate (and government) interests to attract new investment and well-paid, highly-educated residents to the downtown core. This type of development process creates a pattern of spatially uneven investment from one neighborhood to the next, with some experiencing rapid gentrification and others enduring sharp decline (Wilson 1989, 1991). Low-income, arguably also environmentally burdened, communities like West Indianapolis on the near west side of downtown, see redevelopment potential dwindle and interest wane. Neighborhoods like West Indianapolis push to spur growth of residential development in particular but are often stifled by the city's stagnant and narrow vision of the area as a continued industrial/commercial complex. In contrast, certain neighborhoods like Martindale-Brightwood see the city government's interest and investment in the area grow. Particularly interesting is Martindale-Brightwood's position immediately adjacent to a neighborhood (Fall Creek Place) which has undergone rapid development and gentrification over the last decade. It would appear 'sustainable development' is pushing Martindale-Brightwood in the same direction.

13.3 Mapping Environmental Justice: Sustainability Indicators in Indianapolis

Here we examine the spatial distribution of environmental hazards in Indianapolis as a whole and in the near-northeast side neighborhood of Martindale-Brightwood. These hazards can be reconceived as 'sustainability indicators' (Agyeman et al. 2002). In particular, Agyeman (2008) supports the use of three variables (two of which are easily visualized) to serve as indicators of sustainability-related efforts and actions on the part of a city government. Those variables being; (1) Mean income level of people in historically disadvantaged communities; (2) Proportion of environmental pollution sources in historically disadvantaged communities with respect to the city's other communities; and, (3) Participation of HDCs and their self-selected reps in decision-making processes (this one being not as transformable to visualized variables).

In visualizing these 'sustainability indicators', the pollution sources selected here are brownfields (abandoned, unused, or underused sites which have yet to be redeveloped due to actual or potential environmental contamination on the site); toxics release inventory sites (reported releases of 650 listed contaminants on/off site); and, industrial waste generators (facilities required to report how much industrial waste they generate per month). Within each discussion of the three environmental hazards (or 'sustainability indicators') across Indianapolis I first produce maps visualizing the particular hazards in relation to various socioeconomic and racial characteristics across Indianapolis. I then assess and discuss associations between socioeconomic status and racial characteristics of the population (by block group) and the presence or absence of the environmental hazards. This includes a discussion of specific environmental hazards endured by residents of Martindale-Brightwood as well as other communities with a relatively high number of environmental issues.

To analyze environmental injustice, ultimately as an indicator of sustainability efforts, a logistic regression analysis was performed at the block group level. The dependent variable is presence or absence of brownfields/toxic release sites/industrial waste sites in the block group, and the independent variables are median household income, percent African-American population, percent female-headed households, percent housing vacancies, and percent of housing units renter-occupied. The logistic regression analysis for brownfields revealed a strong and statistically significant relationship between brownfields and every socioeconomic factor, including percent African-American population (Table 13.1).

This result illustrates the oft-cited burden of housing brownfields in communities which have historically endured disinvestment. Brownfields are the only environmental hazard assessed here with a strong significant correlation with race (% African-American). The abundance of brownfields within communities like Martindale-Brightwood serve as not only scars of industrial decline, but potential wounds from an environmental/human health perspective. This finding is not surprising if one is familiar at all with the environmental justice movement and literature in the U.S.

The results listed in Table 13.1 reveal the strong relationships brownfields hold with not only race but also with median household income (Fig. 13.2), vacant housing, and renters. Brownfields therefore are significantly associated with a host of social inequities. Because environmental 'cleanup' of these sites only occurs after the damaged property is purchased or targeted for redevelopment, cleanup is wholly dependent upon the whims of capital. If no investment interest is generated for a brownfield site, then that site will remain a hazard indefinitely. In this way, race and class are inculcated within any discussion of environmental injustice in Indianapolis, Indiana. Based on data from the 2010 Census, the median household income in block groups containing at least one brownfield is \$28,725 compared to \$42,144 in those without brownfields.

Table 13.1 Logistic binary regression results for brownfields

	Brownfields	
Sociodemographic characteristics	B	Sig.
Median household income (log)	−1.939	**0.000***
African-American (%)	0.948	**0.005***
Female headed household (%)	3.910	**0.008***
Vacant households (%)	3.434	**0.000***
Renter occupied (%)	2.067	**0.000***

**p* significant at the 0.05 level

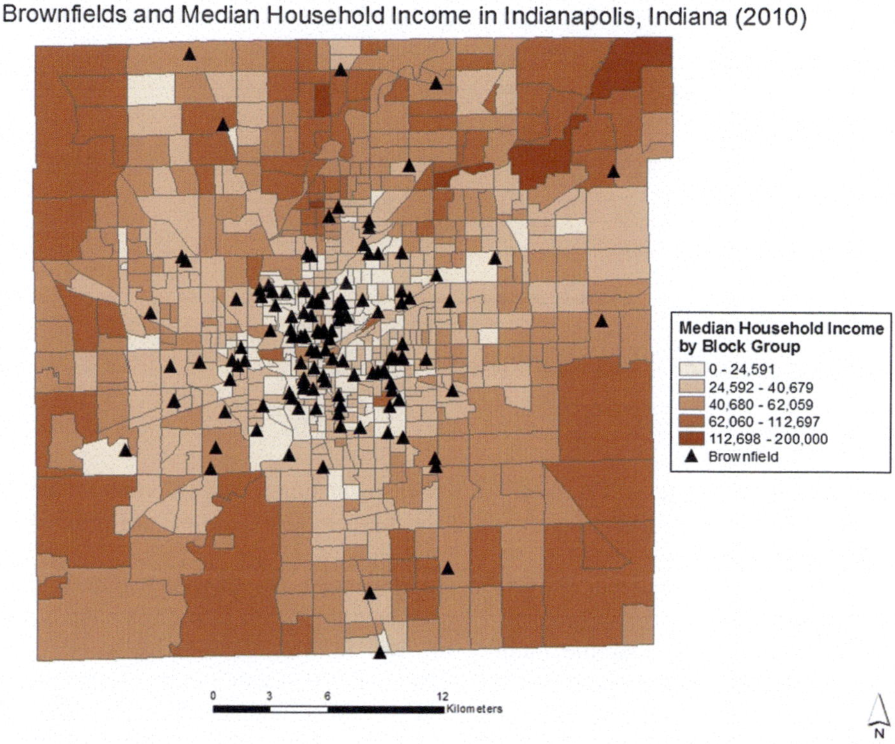

Fig. 13.2 Location of brownfields and the distribution of median household income

13.3.1 Toxics Release Inventory

Toxics release inventory (TRI) sites are recorded releases/leaks at facilities. Data about these releases are maintained by the EPA for monitoring, regulation and enforcement purposes. The data on TRI sites used in this research project were acquired from the U.S. EPA for the year 2013. Over the course of 20 years (1989–2009) the total number of TRI sites in Indianapolis decreased from 103 to 77 (U.S. EPA 2010). However, the total amount of materials released (both on and off-site) actually increased from 9,138,045 to 10,922,519 lbs. This indicates that while

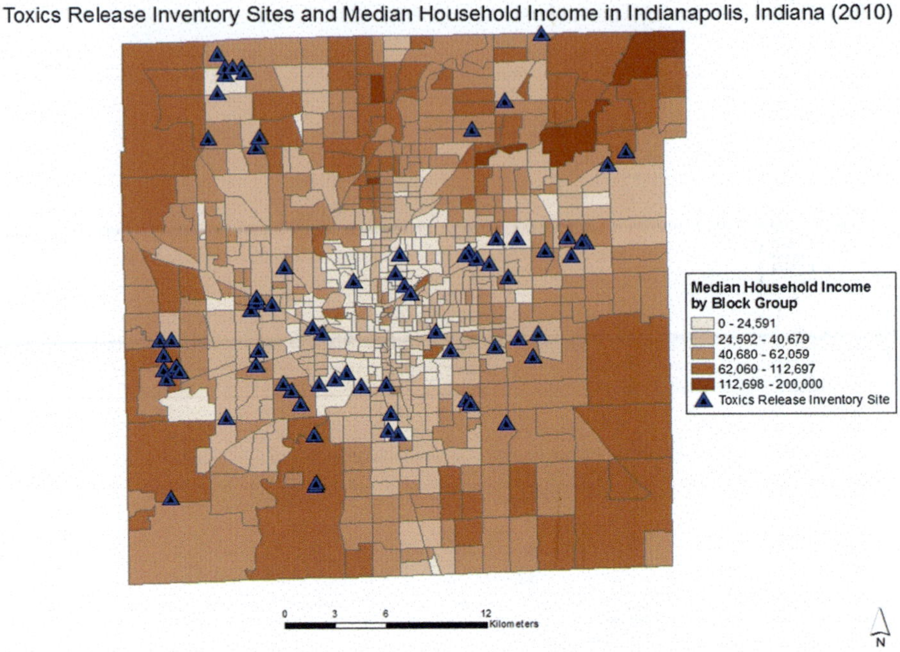

Fig. 13.3 TRI sites (2013) and median household income

the total number of hazardous sites declined, suggesting a positive environmental trend, the increase in total emissions defeats any thought of environmental improvement.

By 2013 the total number of TRI sites in Martindale-Brightwood had decreased further to 74 sites with a total amount released being just over 11 million pounds. Of particular interest is that the number of sites in Martindale-Brightwood only decreased from eight to seven (between 2000 and 2013) while in other communities, such as the low-income white community of West Indianapolis, the number of TRI sites decreased from 16 in 2000 to seven in 2013 (USEPA TRI Explorer). Although TRI sites are unevenly distributed across Indianapolis (Fig. 13.3), statistical analyses show that the distribution is based predominately on income levels (Fig. 13.3) rather than race. Racial characteristics (as of 2010 see Table 13.3) of block group populations have little association with TRI location. Lower income block groups are more likely to contain a TRI site than higher income block groups (Table 13.2). Median household income (log) is significantly and inversely associated with the presence of a TRI site, indicating that the chance of housing a TRI site decreases as block group income increases.

No other independent variables possessed statistically significant associations with TRI location. This illustrates the strong income-based disparity in Indianapolis of the presence of TRI facilities.

Table 13.2 Logistic binary regression results for toxics release inventory sites

	Toxic release inventory	
Sociodemographic characteristics	B	Sig.
Median household income (log)	−1.107	**0.039***
African-American (%)	−0.460	0.333
Female headed household (%)	−0.001	0.826
Vacant households (%)	0.659	0.443
Renter occupied (%)	−0.008	0.989

*p significant at the 0.05 level

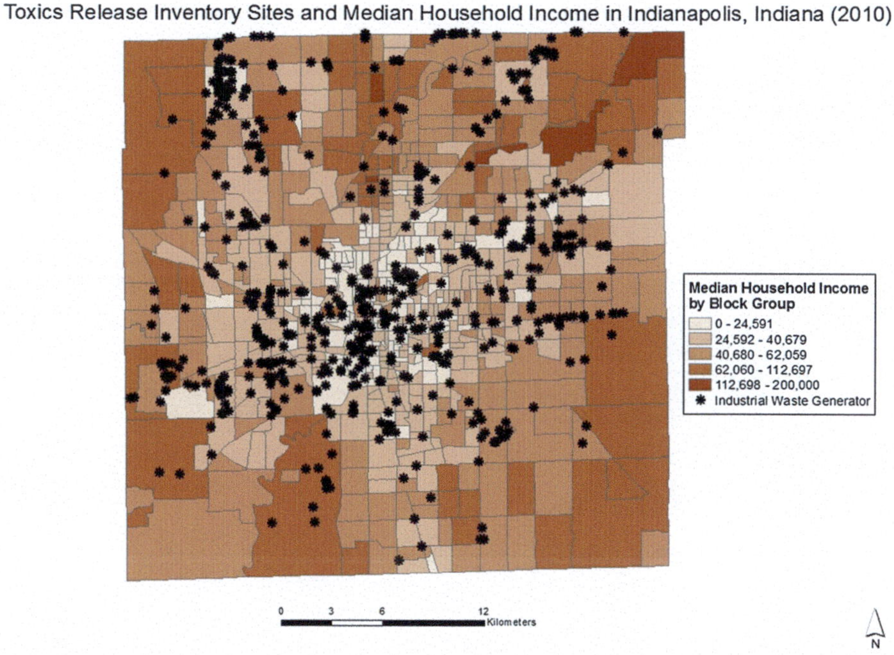

Fig. 13.4 TRI sites (2013) and median household income

13.3.2 Industrial Waste Generators

The Indiana Department of Environmental Management maintains data regarding all facilities within the state of Indiana that generate industrial waste above particular quantities (Fig. 13.4). Here I focus on large quantity generators of industrial waste (LQG), those that generate over 1,000 kg per month (U.S. EPA RCRA). As with the TRI sites while race is not significantly related, income is indeed related to the presence of industrial waste generators. Logistic regression analysis (Table 13.3) revealed a significant negative relationship ($b=-1.008$; $p=0.009$) between industrial waste generators and median household income.

Table 13.3 Logistic regression results for industrial waste generators

	Industrial waste generators	
Sociodemographic characteristics	B	Sig.
Median household income (log)	−1.008	**0.009***
African-American (%)	−0.105	0.663
Female headed household (%)	0.385	0.735
Vacant households (%)	1.313	**0.040***
Renter occupied (%)	1.188	**0.000***

*p significant at the 0.05 level

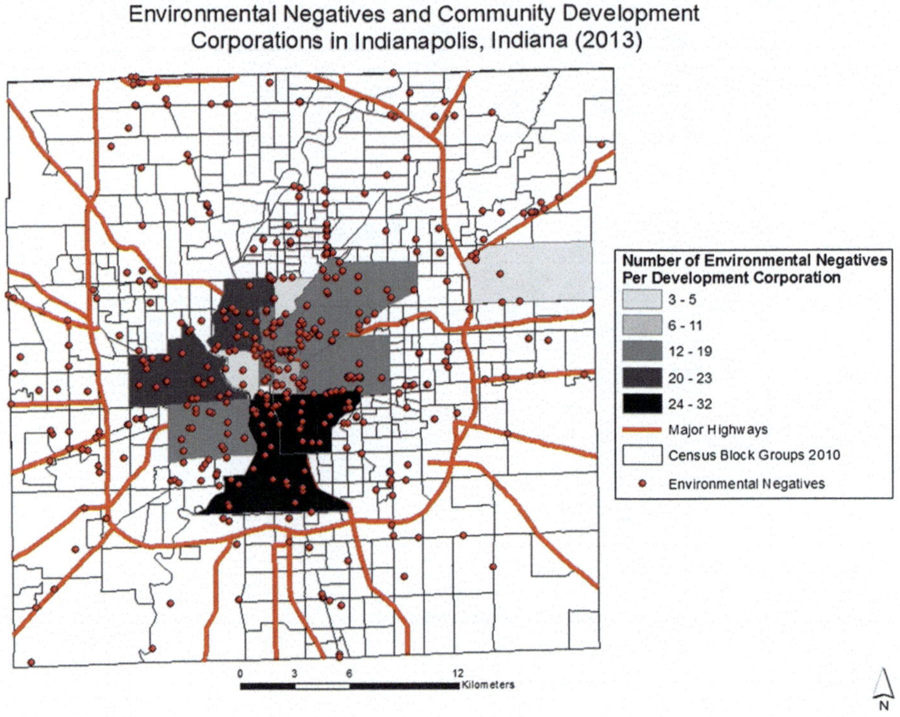

Fig. 13.5 Environmental negatives and community development corporations

As income decreases in a block group, the likelihood of finding an industrial waste generator in that block group increases. In addition, a statistically significant positive relationship was discovered between industrial waste generator sites and vacant households ($p=0.040$) as well as renter-occupied homes ($p=0.000$). As the number of vacant households and/or renter-occupied homes increases in Indianapolis, so too does the presence of industrial waste generator sites. Median household income is the strongest predictor for all of the environmental hazards included in this analysis.

However, race is significantly associated with only brownfields. These results illustrate the strong links between socially and economically disadvantaged residents and the presence of these particular environmental hazards. The neighborhood targeted as part of Indianapolis' "Smart Growth Redevelopment District", Martindale-Brightwood, has long been a socially, politically, economically, and environmentally depleted landscape (Fig. 13.5). However, it should be noted that Martindale-Brightwood, the site of the city's 'Sustainable Community' efforts, was actually the 8th out of 14 community development corporations when ranked according to the number of environmental negatives found within each CDC's boundary.

Following many years of blatant and subtle racist practices on the part of the City of Indianapolis and its more powerful residents, decades of disinvestment and sociopolitical exclusion produced an environmental 'hazardscape'. Disadvantage exists not simply in the presence of environmental hazards, but also in the political and economic processes that underpin such hazards. In this way, sustainability proponents call for a broader remedy than purely 'green' measures (Agyeman et al. 2002). Long-term disinvestment both economic, with the retreat and relocation of various industrial operations, and social, in the strategic retreat of social services has dramatically shaped the landscape of Martindale-Brightwood (Wilson 2007). Martindale-Brightwood serves as a reminder of long-term racial politics resulting in the isolation of communities like it.

13.4 Conclusion

The purpose of this research was to examine the efforts of the City of Indianapolis, Indiana to create a 'sustainable community'. The current spatial distribution of environmental hazards is predominantly based on class or income differences across the city of Indianapolis. Environmental injustice in Indianapolis is rooted in class-based privileges as well as race (with respect to brownfields). The concentration of brownfields in MB is also a symbol of long-term disinvestment, as brownfields are legacies of the area's former industrial base. The City of Indianapolis Brownfields Coordinator recognized this overwhelming concentration of brownfields within the Martindale-Brightwood community and sought Federal funding to create what became the 'Smart Growth Redevelopment District' and the Brownfield Pilot effort. The city has played a role in several significant brownfields redevelopment success stories over the last few years. This research does not question the intent of the city in redeveloping this area. Brownfields redevelopment must be a component in the pursuit of a 'sustainable community'. Sustainability as an approach to urban planning and redevelopment must not only be implemented in a broader fashion but, should also expand its framework to include issues of social justice rather than simplistic economic redevelopment mechanisms that often provide little to no direct benefit to nearby residents.

The City of Indianapolis leadership has envisioned the future of Martindale-Brightwood (underutilized property, ripe for injection of capital using brownfields and sustainability). Connections to local and federal government agencies in MB, while beneficial to MB residents in one sense, fostered the eventual co-optation of residents' vision of change by city and federal agencies. The risk of a redevelopment approach rooted in economic conceptions of sustainability as opposed to social and environmental conceptions will likely produce a narrowly 'sustainable' but broadly gentrified community. While the city's efforts have been productive in altering certain environmental hazards in the pilot project community (MB), it must be careful not to exclude residents from decision-making processes. The city must make great strides in integrating and explicitly discussing the issues which run deeper than an environmental remediation, those being concern for social justice and economic equity as components of a sustainability project. All three components must be equally and concurrently pursued – environmental equity = economic equity = social equity. The current approach of overly focused economic development will simply produce a new form of inequitable distribution manifested in a newly gentrified neighborhood.

References

Agyeman J (2005) Sustainable communities and the challenge of environmental justice. New York University Press, New York

Agyeman J (2008) Toward a 'just' sustainability? Contin J Media Cult Stud 22(6):751–756

Agyeman J, Bullard RD, Evans B (2002) Exploring the nexus: bringing together sustainability. Environmental justice and equity. Space Polity 6(1):77–90

Alston D (1991) Speech delivered at the First National People of Color Environmental Leadership Summit. Washington, DC, October

Anderton DL, Anderson A, Oakes J, Fraser M (1994) Environmental equity: the demographics of dumping. Demography 31:229–248

Bullard RD (1990) Dumping in Dixie: race, class, and environmental quality. Westview Press, Boulder

Chambers S (2007) Minority empowerment and environmental justice. Urban Aff Rev 43(1):28–54

Checker M (2008) Bringing 'Green Collar' jobs to the South Bronx. Gotham Gazette. August 19. www.gothamgazette.com/article/communitydevelopment/20080819/20/2616

Fuller, Trevor (2014) What prompts activism in response to environmental injustice? Environmental Justice 7(1): 27-32

Hulse LJ, Zeigler C (1991) The suburbanization of Indianapolis: an outline of metropolitan development in Marion County, 1830–1980. POLIS Research Center, Indianapolis

Indianapolis Department of Metropolitan Development (2009) Division of Brownfields Presentation, MBEJC meeting, 2009

Lee C (1992) Toxic waste and race in the United States. In: Bryant B, Mohai P (eds) Race and the incidence of environmental hazards: a time for discourse. Westview, Boulder

Lubitow A, Miller TR (2013) Contesting sustainability: bikes, race, and politics in Portlandia. Environ Just 6(4):121–126

Martindale-Brightwood Community History Profile (2000) The POLIS Center. Indiana University at Indianapolis

Mohai P, Bryant B (eds) (1992) Race and the incidence of environmental hazards: a time for discourse. Westview, Boulder
Pulido L (2000) Rethinking environmental racism: white privilege and urban development in southern California. Ann Assoc Am Geogr 90:12–40
SAR International, Inc (1987) Partnership for sustainable communities EPA brownfield pilot – Indianapolis, IN. September 2011. United Church of Christ Commission for Racial Justice. Toxic wastes and race in the United States: a National Report on the Racial and Socioeconomic Characteristics of Communities with Hazardous Waste Sites. Public Data Access, New York
United Church of Christ Commission for Racial Justice (1987) Toxic wastes and race in the United States: A national report on the racial and socioeconomic characteristics of communities with hazardous waste sites. Public Data Access, New York
USEPA, HUD, DOT (2010) Partnership for sustainable communities: a year of professional development for American communities, October
Warner K (2002) Linking local sustainability initiatives with environ- mental justice. Local Environ 7(1):35–47
Wilson D (1989) Local state dynamics and gentrification in Indianapolis, Indiana. Urban Geogr 10:19–40
Wilson D (1991) Urban change, circuits of capital, and uneven development. Prof Geogr 43(4):403–415
Wilson D (2007) City transformation and the global trope: Indianapolis and Cleveland. Globalizations 4(1):29–44

Chapter 14
Unmanned Systems and Managing from Above: The Practical Implications of UAVs for Research Applications Addressing Urban Sustainability

Karen Gallagher and Patrick Lawrence

Abstract The development and widespread use of small Unmanned Aerial Vehicles (UAVs), such as quadcopters with GoPro cameras, is leading a revolution in a wide range of hobby, commercial and potential research applications of the technology. Within urban areas the variety and pace of changes occurring with both human and natural systems have significant impacts on efforts to address and manage sustainability. A wide range of opportunities and challenges exist in considering how UAVs may effectively be used as a spatial tool in examining urban areas, including buildings, infrastructure, ecosystem features and processes, natural areas, and environmental health. This chapter examines the technical, practical, policy and legal implications of the use of UAVs to assess and monitor urban sustainability and includes highlighting examples of recent research applications. Uses of scale, resolution, accuracy and reliability are noted along with how improving technologies and the growth in the widespread familiarity and use of UAVs represent an expanding field for urban research applications. Challenges including privacy, federal regulations governing the commercial use of UAVs are discussed. A series of recommendation research applications are presented in terms of the greatest potential where UAVs can be employed to further research, understanding, and decision-making in the area of urban sustainability.

Keywords Urban • Environment • Drones

14.1 Introduction

Innovations in geospatial science and remote sensing have enhanced our understanding of the built and natural environment by allowing for data collection that covers large areas of the earth's surface from the atmosphere. Advancements beginning in

K. Gallagher (✉) • P. Lawrence
Department of Geography and Planning, University of Toledo, Toledo, OH, USA
e-mail: karen.gallagher@rockets.utoledo.edu; patrick.lawrence@utoledo.edu

J.D. Gatrell et al. (eds.), *Urban Sustainability: Policy and Praxis*,
Geotechnologies and the Environment 14, DOI 10.1007/978-3-319-26218-5_14

the seventeenth century have resulted in a capacity to collect information from a range of areas and resolutions using frequencies of electromagnetic radiation that is reflected from surfaces back to recording platforms. Improvements in photography and methods used to mount and utilize airborne photographic equipment have also increased the efficiency and accuracy with which it is possible to collect data. The most common modes of collecting remotely sensed data – the flight of manned aircraft and from orbiting satellites – have left a need for the development of a technology that can amass images at a higher resolution that provides a more detailed representation of the surface of the earth. The gap in the ability to collect spatial data lies in a portion of the atmosphere where aircraft cannot operate and the resolution from satellite imagery is not sufficient to monitor environmental processes.

Unmanned aerial vehicles (UAV) – remotely controlled aircrafts that do not require a flight operator to be on board – (also often referred to as drones) have the potential to collect data at a scale and resolution that has not been possible with modern technology. Typical types of smaller UAVs that are used for hobby photography or filming and other civilian and potential commercial and research applications include small planes and helicopter frames often only a few feet in size. Quadcopters with four, six or eight rotter blades are becoming very common as costs have declined in recent years (to under $400 per base unit), their widespread availability from vendors, relative ease of operation, use of GoPro cameras, and the development of accessory equipment to support mounting of cameras and stabilization of the platforms while in flight (Fig. 14.1). IGI Consulting Inc. estimates that the U.S. market for small commercial and hobby UAVs will grow from $5 billion in 2013 to $15 billion in 2020.

Within the last few years, these types of small UAVs have presented new opportunities and challenges for scientists who examine changes to natural and human characteristics of the earth's surface and contribute to a better understanding of a

Fig. 14.1 Quadcopter drone phantom (Photo by DJI Ltd.)

wide range of issues, including urban sustainability. The low cost of acquiring and operating UAVs – along with the ability to conduct autonomous missions – offers an attractive tool for the collection of near surface remotely sensed data. Conducting studies that employ UAVs can also reduce the cost of data collection because they can provide a safe and secure means to conduct research. Currently the use and application of UAVs is limited due to federal regulations concerning airspace, public safety, and privacy rights (Federal Aviation Administration 2015b). The Federal Aviation Administration (FAA) (2015b) stipulates that UAVs must not exceed an altitude of 400 ft and must stay within sight of the ground control operator. UAVs are not to be flown in populated areas, those hoping to conduct research in densely populated areas may apply for a special airworthiness certificate which can take anywhere from 2 months to 1 year for approval (Federal Aviation Administration 2015b). If an airworthiness certificate is acquired it is stipulated that the UAV operator must have a FAA-issued pilot certificate in order to fly the mission. In February 2015 the FAA also released proposed guidelines for the commercial use of UAVs that may result in licensed applications by permitted private sector companies, which may impact the ability for their use in many potential research applications.

However, the limitations placed on UAVs have left enough leeway to test their exploratory application in environmental research. Though examples are limited, those who have conducted studies with UAVs were able to fly the aircraft in the operator's line of sight and follow other safety and operational standards. The studies were also conducted in areas of low population density or above the owner/operator's personal property (Whitehead et al. 2014). To this point, case studies have uncovered a need for development in remote sensing technologies such as lighter cameras that can collect images at a higher speed and resolution. This is mainly because high resolution cameras are either too heavy or too large to safely adhere to the UAV platform. If advanced camera technologies were made commercially available – and at an affordable cost – it would increase the sophistication with which remotely sensed data could be collected when using UAVs. There is also a need for UAVs with a longer battery life so that longer missions can be flown. Though UAVs have presented opportunities for environmental research, at present numerous regulatory and technological limitations preclude many of the possibilities. However, examining past experiences with collecting remotely sensed images can demonstrate how the technology and its applications have evolved and can provide examples of how technological advancements can continue to grow the state of opportunity in using UAVs to better understand urban environments.

The earliest attempts at remotely sensed data collection began in the mid 1800s when cameras were mounted to airborne objects such as birds, kites, and hot air balloons. The result was the depiction of still frame images that portrayed the composition of cities and the aftermath of natural disasters (Bhatta 2009). Though the ability to collect aerial imagery provided a new way to observe the earth's surface, the quality of images and the amount of land that could be observed was limited by technology. The invention of the airplane in the early 1900s made it possible to collect data over much larger spaces. In combination with advances in photography, aviation allowed for the collection of aerial images that could portray water depths and vegetation indicators, which advanced military planning during WWII (Bhatta 2009). Further advances during the space age in the 1950s and 1960s made way for data collection from above the earth's atmosphere. Cameras were mounted to rockets and satellites with the ability to collect data for large swaths of the earth were launched into orbit. At the same time, camera technology was improved and new methods of collecting images using infrared and multi-spectral data collection techniques were developed.

The first use of UAVs can be traced back to WWI with the deployment of aerial torpedoes (Keane and Carr 2013). Since then, United States military agencies have invested in further developing the capabilities of UAVs for use in combat and for the collection of intelligence data. By the Gulf War of the early 1990s, UAVs were reliable enough to use for reconnaissance missions that provided military personnel with real-time intelligence data about enemy troops (Keane and Carr 2013). Further developments in UAV technology have resulted in systems that are competitively priced when compared to piloted flight missions. The low price and autonomy offered by UAVs have made them desirable on the commercial market. Despite regulations that restrict the use of UAVs their applications for research and industry have been explored.

14.2 Applications

In recent years, numerous case studies have been conducted to explore the role of UAVs in scientific research. These are primarily focused in the areas of wildlife monitoring, habitat surveying, mapping, and topographic modeling (Chabot and Bird 2013; Mancini et al. 2013; Whitehead et al. 2014). In addition to exploring application, researchers have begun to conceptualize how UAV's can be used in climate research, disaster relief, natural hazard assessment, and atmospheric monitoring (Ajibola and Mansor 2013; Giles 2014; Xie et al. 2013). However, current (as of February 2015) federal policy limits the use of UAVs for scientific research and environmental monitoring. Of particular interest are the examples of – and potential for – the use of UAVs in examining, monitoring and assessing conditions within urban environments including the examination of water resources, the urban heat island effect, green spaces, urban wildlife populations, transportation, infrastructure and land use planning (Barnes et al. 2014; Dominici et al. 2014; Gruen 2013; Haarbrink and Eisenbeiss 2008; Ro et al. 2007; Tahar 2012).

Of note this discussion will focus on applications of small commercial UAVs for research addressing aspects of urban sustainability – not military or surveillance uses. Whitehead et al. (2014) discuss the potential uses of UAVs in cities when the cost, resolution and operational issues of more conventional remote sensing – via planes or satellites – are limited or not feasible. Yuhas (2006) provides a comprehensive review of the development of UAVs and use in studies ranging from atmospheric conditions, air pollution and land management applications. O'Callaghan (2014) notes UAVs as one of the five digital trends for planners in 2015. Schoellig

(2014) presents an assessment of the broader range of applications for UAVs in future urban environments, including monitoring of infrastructure, 3D mapping, and powerline/pipeline inspections, and outlines the advantages of UAV technologies: provides an on demand 'sensor in the sky' readily deployed as needed, provides high resolution temporal and spatial data, and able to execute tasks repetitively, fast, and accurately.

An emerging use of UAVs is in the area of wildlife monitoring. Studies that observe animals in their natural habitats have been conducted in various land and aquatic environments. For example, a UAV was employed by researchers hoping to monitor a salmon spawning event along a 1 km stretch of a river in British Columbia. The researchers flew a UAV 150 m above ground which allowed for imagery collection at an approximate resolution of 5 cm. With this, they were able to determine locations of major spawning concentrations and to locate individual salmon in the spawning grounds (Whitehead et al. 2014). A second study conducted by Sarda-Palomera et al. (2012) collected fine scale aerial images depicting the nesting distribution of the black headed gull in Cantalonia, Spain. The use of UAV for monitoring was beneficial because it made data collection possible while minimizing disturbance of the gulls. UAVs have also been used to locate larger mammals using thermal infrared cameras. Whitehead et al. (2014) successfully monitored cattle by detecting the difference in ground temperature and the temperature of the animals. These studies present potential applications for the study of wildlife in urban areas. By using UAVs to monitor wildlife location and migration patterns researchers can better understand how animals move through urban environments. This can also be beneficial for pest control and for considering land uses as it relates to wildlife habitat and migration.

UAVs have demonstrated an effective means and low cost alternative for monitoring the health of agricultural crops through precision agriculture. Precision agriculture typically uses satellite remote sensing or manned flights to identify variability within crops. However, the high cost of chartering manned flights or poor resolution with satellite images can pose challenges for farmers wishing to monitor crops with precision agriculture. This has led some researchers to assess the extent to which UAVs can be used to survey crops and provide farmers with information to help them determine areas of a field that need attention. For instance, Whitehead et al. (2014) was able identify blight using a UAV with a visible and near infrared camera secured to the UAV's platform. Findings from the UAV survey were shared with the farmer who was able to apply fungicide to the affected area and thus contain the blight (Whitehead et al. 2014). UAVs can also be used to identify the water variability of crops by attaching thermal and multispectral cameras to the UAV platform. Baluja et al. (2012) monitored the hydration variability of a vineyard using a UAV. The data was used to create a map that showed the spatial variability of hydration within the crops. The information allowed for the targeted allocation of water resources to where they were needed most.

UAVs have been used to survey lands for a range of purposes including the assessment of vegetation cover, including plant life, forests, and natural habitat conditions. Chabot and Bird (2013) found UAVs to be a convenient, cost effective, and efficient means of assessing habitat in a wetland environment. The researchers were able to survey a 128 ha wetland in order to evaluate the quality of habitat for the *Least bittern*, an avian species deemed threatened by the Canadian Wildlife Service. Though the UAV findings were verified by ground survey (87.19 % kappa coefficient), the images could only be classified into three categories using supervised classification methods: cat tail, other vegetation, and water. To collect greater detail a high resolution camera would have been needed. However, it was not possible to use a high resolution camera because advanced camera technologies are currently not designed to work well on UAV platforms. Despite the limited detail of data collected, the three classes of vegetation cover did provide enough information to inform the creation of a habitat management plan. UAVs have also been used to assess the risk of birds who build their nests on or near power lines. For instance, Mulero-Pázmány et al. (2014) mounted a Go Pro Camera to an UAV and were able to monitor habitat on power lines and pylons while minimizing the risk that human surveyors would normally encounter. The methods used in that study could be extended to assess the risk to other animals who build their homes in potentially dangerous situations such as on smokestacks or near harmful ventilation systems. The UAV allows for up-close views while minimizing harm to people and disturbance to wildlife.

Researchers have begun to explore the effectiveness of UAVs with mounted Lidar scanners for measuring and mapping forests below the canopy (Chisholm et al. 2013). The measurements were field verified and found to be accurate for trees taller than 3 mm and within 3 m of the flight path. Methods from this study could be applied to urban environments that are monitoring the restoration of previously forested patches to their natural forms. Recent urban stream restoration projects in the Toledo OH area have employed UAVs as means of establishing monitoring benchmarks to assist with long term assessment of the habitat improvements in river, wetland and riparian conditions following project construction. At the Toledo Botanical Gardens, pond and river systems have been imaged so future changes to vegetation and habitat conditions can be determined (Fig. 14.2) At the University of Toledo, UAV still photography and video has been recorded as a means of documenting recent aquatic and river bank rehabilitation efforts (Fig. 14.3).

Studies in a number of settings have been conducted to explore the capabilities of using UAVs combined with LIDAR scanners for the purpose of creating topographic or digital elevation (DEM) models. Such models have been constructed for the purpose of monitoring coastal erosion, landslides, and land surface changes through time (Mancini et al. 2013; Whitehead et al. 2014). UAVs have also been used to map flood plains by combining terrestrial laser scanning with UAV based bathymetry modeling (Flener et al. 2013). Researchers in the aforementioned studies verified their findings using geospatially referenced ground control points and found a high level of quality and vertical accuracy in all cases. Challenges cited included difficulty executing the flight in windy conditions and issues with imagery due to the reflection of sunlight from the water, but the researchers found their UAV studies to be efficient and cost effective. Due to the ease of operating the UAV they were also able to collect data for areas that were previously inaccessible (Whitehead

Fig. 14.2 Aerial UAV image of the Toledo Botanical Garden restoration project (Image provided by Matt Horvat, TMACOG)

Fig. 14.3 Stream bank along Ottawa River, University of Toledo during UAV filming (Image provided by Patrick Lawrence, photography by Rick Francis)

et al. 2014). The ease of data collection and the frequency with which it can be conducted at a low cost also make UAVs a prime tool for assessing landscape changes that occur slowly over time or quickly as a result of natural disasters. This could be particularly useful in preventing the loss of homes and combatting erosive forces in areas with a high population density.

Scientists have begun to explore the potential for atmospheric monitoring using UAVs. However, researchers have not been able to fully explore the method due to the current policy restrictions imposed by the federal government. As of 2014, the only agencies authorized to use drones for scientific research are NASA and NOAA (Giles 2014). Revised federal regulations and guidelines would allow more researchers to use UAVs for real time data collection that could permit accurate and quick weather forecasting, including the forecasting of weather for urban areas. Currently the best practice is using air balloons which cannot cover as large of an area as UAVs and are not effective in collecting data from the section of the atmosphere that is difficult for satellites to reach (Giles 2014). The federal policy limitations also affect the use of UAVs in disaster relief efforts. An example of this occurred in Boulder, CO where a private company was donating their time and service by deploying drones to assess flood damage in areas that were difficult for manned aircrafts to monitor. However, the company was ordered to stop the mission and threatened with prosecution once the Federal Emergency Management Agency (FEMA) arrived on the scene (Ackerman 2013).

Prior to 2007 when the FAA imposed a nationwide no-fly zone for the commercial use of UAVs, companies were starting to use them to monitor defects in pipelines, power lines, wind turbines, smokestacks and other structures that would pose a danger to human surveyors (Woody 2014). UAVs – in combination with heat sensor technology – have also been used to monitor the energy efficiency of buildings by looking for temperature changes along windows on the outside of buildings (Martinez-De Dios and Ollero 2006). UAVs have successfully been deployed to survey the condition of roadways and bridges. Conducting a primary survey with UAVs keeps transportation workers out of heavy traffic which can be dangerous and at times fatal to on the ground surveyors. Researchers at Michigan Tech University are currently studying the potential of UAVs to assess the conditions and identify priority rehabilitation needs for highway bridges, road surfaces, and geotechnical studies for land slope hazards threatening infrastructure including railway lines (Fig. 14.4) (Brooks et al. 2014; Ahlborn et al. 2013).

14.3 Discussion

Though the National Transportation Safety Board ordered the ban on commercial drones to be lifted in March of 2014 the proposal for new regulations was not released by the Federal Aviation Administration (FAA) until February of 2015 (Jansen 2015). Once approved, the FAA will regulate the extent to which UAVs can be used and set conditions that UAV operators must meet. For instance, in the

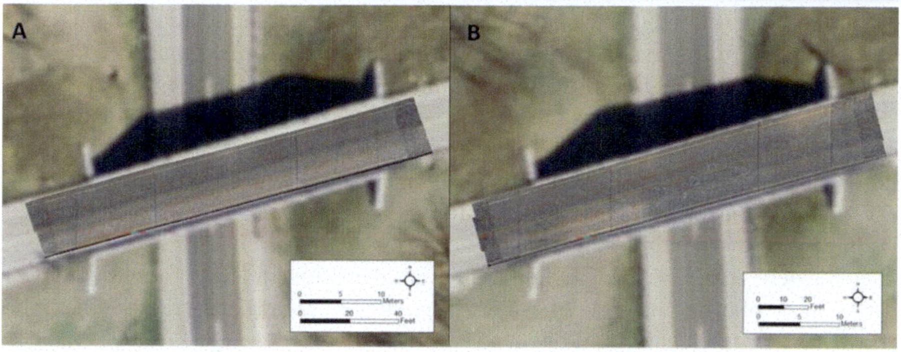

Fig. 14.4 Example of UAV imagery used to evaluation bridge conditions (Photo courtesy of Michigan Tech, see Ahlborn et al. 2013)

proposal the FAA (2015a) stipulates that UAVs must be flown during daylight hours, no higher than 500 ft and must not exceed a speed of 100 mph. Furthermore, the total weight of the UAV system must not exceed 55 lbs. The flight operator must be at least 17 years of age, have passed an FAA approved aeronautics test to obtain a private pilot's license and clear the necessary background checks conducted by the Transportation Security Administration (Federal Aviation Administration 2015a).

Although limiting, the FAAs new regulations will make the way for more commercial use of UAVs. A report released by Lux Research predicts that the commercial market for UAVs will reach $1.7 billion by 2025 (Jacques 2014). The agriculture industry is expected to lead the market demand for UAV services, which is likely a result of the potential for UAVs to provide a low cost means of allocating resources through precision agriculture (Baluja et al. 2012; Jacques 2014; Whitehead et al. 2014). The commercial drive to develop uses of UAVs for agriculture could provide an increased need for studying the ways in which UAVs can benefit crop yield.

It is also expected that commercial services in operation before the nationwide ban enacted in 2007 will once again emerge. Lifting the ban will allow businesses that used UAVs to monitor pipelines, smokestacks, and other potentially hazardous features to resume services (Linkhorn 2015; Woody 2014). There is great potential to expand the potential range of settings in which UAVs can be used to conduct assessments in hazardous situations. The exploration is not just limited to the use of UAVs for surveying damaged infrastructure, but can also extend to identifying the impacts to wildlife that nest on power lines and other potentially dangerous situations (Mulero-Pázmány et al. 2014). As the technology develops further and becomes more accessible and cost effective, it is possible that UAVs could become part of an emergency monitoring system that uses real-time notification to alarm personnel when noxious gasses are released into the atmosphere (Xie et al. 2013). However, this idea has only been conceptualized as it is thought that interactive data delivery and processing systems that fit on a UAV platform must be developed before the real-time monitoring system can become a reality (Xie et al. 2013).

Further advancements in camera technologies leading to lighter and smaller high resolution equipment available on the commercial market can also increase the possibility of UAVs to be used for habitat monitoring. Though research by Chabot and Bird (2013) was able to distinguish three land cover classes in a wetland environment the researchers would have collected more detailed information if a high resolution camera would have adhered to the UAV platform. UAVs are beneficial for the use of classifying and monitoring vegetation because of their ease of use and for the frequency with which missions can be flown at a low cost. Studies have been conducted where UAVs were used to monitor coastal erosion and have been found to be useful in combatting the loss of homes in populated areas that are exposed to the corrosive forces of nature (Whitehead et al. 2014). However, to further enhance capabilities of UAVs, work could be done to increase the reliability and consistency of UAVs and cameras when collecting data in high winds, varying light and inclement conditions.

A new area of research lies with exploring the application of UAVs in many other aspects of urban sustainability research. To this point, many uses have been considered, including mapping landscapes, monitoring urban land cover change, surveying crops for precision agriculture, surveying wildlife and habitat, and using UAVs to assess potentially hazardous situations. In the coming years further advances in technology and a less restrictive policy could increase potential uses of UAVs. To date UAVs have proven to be effective in collecting data at a detailed resolution from an area of the atmosphere that has been difficult to reach. Penetrating the section of the atmosphere where manned aircraft cannot fly and obtaining a level of resolution not provided by satellites has presented an exciting potential for the use of UAVs. Though exploration is needed to determine how UAVs can be used for environmental research one thing is certain, a great potential lies with using UAVs to better understand the urban environment.

An important additional issue to consider in regards to UAVs and their application for research in urban areas involves privacy and security concerns. In recent years questions have been raised during the rapid use and deployment of small UAVs in urban areas with mounted cameras and the potential for invasion of privacy with the images capturing persons and private property. In addition, heighted concerns regarding security are increasingly becoming a factor as more small UAVs appear in urban areas, especially those areas with sensitive political, industrial and military targets. The threat comes with the ability of a UAV to record classified information and with its potential to carry explosive devices. Incidents in early 2015 in Paris and Washington DC highlighted that such concerns may be warranted. As the use and appearance of UAVs increase in urban areas, and in response to revised FAA regulations regarding their use, possible impacts on the ability to engage UAVs for research in urban areas may be constrained and certainly questioned. As the discussions centered around the advantages and potential uses and expanded roles for UAVs in studying issues and aspects of urban sustainability continue addressing potential privacy and security concerns will need to be addressed but should not necessarily be viewed as absolute certain barriers to their applicability. For more discussion of the variety of legal issues associated with use of UAVs the reader is referred to Dolan and Thompson (2013).

Based on a review and evaluation of a wide range of applications of UAVs, and considering emerging opportunities with revised federal regulations dictating their operation and likely advances in the technologies with UAV equipment, cameras, GPS, software, related spatial and processing capabilities, we can postulate on the emergence of several applications in the field of urban sustainability and the documenting, assessment, and monitoring of natural and human aspects of urban systems:

1. Green space and Open Space Planning: Including the mapping and evaluation of habitat and other natural characteristics of parks, protected areas, natural areas and corridors, river and riparian, wetland and other areas with urban landscapes, in order to assist with planning and decision-making of land uses and potential re-uses of abandoned urban areas within cities.
2. Urban Forests: Inventory and studies of health of urban trees in order to determine stresses, evidence of infestations, disease, and need for replacement or the addition of new trees in support of the important role of forests in urban systems for shade, microclimate conditions, aesthetics, health, air quality and overall improved environmental conditions (Fig. 14.5).
3. Infrastructure Assessment: Examination and evaluation of the existing conditions of roads, railways, bridges, buildings, surface utility features (power lines), and other built items in order to document degradation and prioritize repairs or replacement, and as a means to address potential public safety and liability concerns (Fig. 14.6).

Fig. 14.5 Monitoring of controlled burn in Oak Openings Savanna forest in NW Ohio (Photo by Rick Francis)

Fig. 14.6 UAV image of former water plant at University of Toledo Medical Campus (Photo by Rick Francis)

4. Gardens and farms: Support the emerging interest and development of urban agriculture through monitoring of site and crop status, assisting with the location and placement of new locations for urban farms, examining soil and drainage conditions, and documentation of urban farming within the context of its role and importance for the urban sustainability movement.
5. Urban rivers and streams: Potential study and monitoring of flood conditions, habitat and overall assessment of environmental concerns and quality of aquatic ecosystems within urban areas where rivers and streams often provide important roles for natural species, recreational opportunities, flood storage capacity following high rain and water flow events, within urban areas.
6. Studies of the Urban Heat Island Effect: Examining the role of differing surfaces, natural and human, in contributing to heat loss and related microclimate conditions within high density urban areas, including tracking the loss of internal heat from buildings, increases in impervious (warm) surfaces, and overall assessment of the contributions of the built human environment to altering local climate conditions.
7. Land Use Planning Applications: Provide more detailed assessment of current and potential future land uses to assist short and long term efforts by local governments in the planning and decision-making associated with land use zoning, official plans, and regional long term strategic efforts at land use planning within urban or urbanizing communities. This could include inventories of the types, amount and locations of land uses (residential, commercial, industrial, open space), plus supporting infrastructure (roads, highways, railways, ports, bridges) to better improve planning for growth and future development, and

Fig. 14.7 Example of typical UAV imagery of urban residential area (Photo by Alan Bunyan)

potential re-use of abandoned urban lands such as brown spaces for alternative uses or the conversion to green spaces (Fig. 14.7).

8. Wildlife Tracking: In urban areas with increased presence of animals, especially those viewed as disruptive or nuisance – such as the rapid expansion of white tailed deer in many cities within the Midwest US – such technologies could increasingly be deployed as a means of using infrared and other sensors to monitor and assess wildlife populations in order to better inform potential actions taken to address problems and determine management strategies.
9. Location and routing of services and materials: Within the context of the rapid changes underway in many urban environments with the redistributions of populations, jobs, commercial and industrial facilities (including the resulting "shrinking city phenomena"), UAV technology – through monitoring and assessing of land uses and human activities could be utilized to help re-design urban areas or entire cities to assist in maximizing the use of existing roads and infrastructure to move goods and people and reallocate land uses and human activities in a more efficient and effective manner resulting in long term cost savings and more sustainable use of land and resources.
10. Potential applications with new sensors: Advances in the range of sensor platforms are underway that would allow for high resolution cover over large urban areas with further range of spectral signatures that may increase the ability of UAVs to detect surface contaminants, water quality conditions, soil characteristics, structural elements of building and other infrastructure, and atmospheric and near ground air characteristics (such as airborne contaminants, smog and ground ozone conditions). These opportunities would provide a great improvement in the assessment of the urban environment and greatly aid in advancing ideas and practices associated with sustainability.

14.4 Conclusions

As discussed, there is great potential for the use of UAVs in urban sustainability research and applications. However, there are current limitations in the form of policy and technology that need to be considered and understood. Though commercial cameras work well, experts agree that high resolution cameras with a compact design will expand the possibilities for studies using UAVs for imagery and data collection. Furthermore, less stringent federal regulations will widen the opportunities for gaining an enhanced understanding of atmospheric conditions and disaster relief efforts which will open up the opportunities for research and commercialized applications. This could be particularly useful due to the ability of UAVs to communicate real time data in urban environments which would allow for more timely and rapid monitoring in situ with detailed coverage over large areas. Despite the challenges, opportunities exist for the use of UAVs in sustainability research as it is carried out in our built and natural environment.

This chapter has been intended to provide for a review and discussion of the current application of UAV technologies that have been – or could potentially be – applied in urban areas within the context of improving our knowledge and management practices aligned with efforts to further develop the concepts and practices associated with sustainability. Several examples of the use of UAVs have been presented in order to illustrate how the technology is currently used in support of – or to direct – research and commercial applications. Major limitations related to the technology (and subsequent need for improved sensors, cameras, software capabilities and platforms) have been outlined, along with the current and future federal regulatory aspects that limit and restrict use and will continue to constrain – to some degree – the practical applications. With the potential and implications in mind, ten applications of UAVs in urban environments to advance sustainability practices have been presented simply as a start to consider how this technology can improve our ability to understand and manage many of the characteristics of complex urban systems, including the vast array of human and natural conditions and factors that make up and alter these areas. As far as the potential for UAVs to enhance our understanding of urban environments it would appear that in many ways it may be considered that the "sky is the limit."

References

Ackerman E (2013) UAV provides Colorado flooding assistance until FEMA freaks out. http://spectrum.ieee.org/automaton/robotics/aerial-robots/falcon-uav-provides-colorado-flooding-assistance-until-fema-freaks-out

Ahlborn TM, Shuckman RS, Sutter LL, Harris DK, Brooks CN, Burns JW (2013) Bridge condition assessment using remote sensors. Michigan Technological University, Ann Arbor

Ajibola II, Mansor SB (2013) UAV-based imaging for environmental sustainability-flash floods control perspective. FIG Working Week 2013. Environment for Sustainability. Abuja, 6–10 May, 2013

Barnes G, Volkmann W, Sherko R, Kelm K (2014) Drones for peace: part 1 of 2: design and testing of UAV-based cadastral surveying and mapping methodology in Albania. Integrating land governance into the post-2015 agenda, harnessing synergies for implementation and monitoring impact. Annual World Bank conference on land and poverty, Washington, DC, 24–27 March 2014

Baluja J, Diago MP, Balda P, Zorer R, Meggio F, Morales F, Tardaguila J (2012) Assessment of vineyard water status variability by thermal and multispectral imagery using an unmanned aerial vehicle (UAV). Irrig Sci 30(6):511–522

Bhatta B (2009) Remote sensing and GIS. Oxford University Press, New York

Brooks C, Roussi C, Colling T, Dobson R, Watkins MK, Dean D, Hart B, Garbarino J, White B, Banach D (2014) Characterization of unpaved road condition through the use of remote sensing project. Michigan Technological University, Ann Arbor

Chabot D, Bird DM (2013) Small unmanned aircraft: precise and convenient new tools for surveying wetlands. J Unmanned Veh Syst 01(01):15–24. doi:10.1139/juvs-2013-0014

Chisholm RA, Cui J, Lum SKY, Chen BM (2013) UAV LiDAR for below-canopy forest surveys. J Unmanned Veh Syst 01(01):61–68. doi:10.1139/juvs-2013-0017

Dolan AM, Thompson RM (2013) Integration of drones into domestic airspace: selected legal issues. CRS report for Congress. Congressional Research Service 7-5700, Washington, DC

Dominici D, Alicandro M, Massimi V, Elaiopoulos M (2014). Surveying with UAV infrared sensors for the urban and eco-sustainable requalification of old cities. FIG Congress 2014, engaging the challenges-enhancing the relevance, Kuala Lumpur, 16–21 June 2014, Paper 7106

Federal Aviation Administration (2015a) Overview of small UAS notice of proposed rulemaking. Washington, DC. Retrieved 22 Feb 2015. http://www.faa.gov/regulations_policies/rulemaking/media/021515_suas_summary.pdf

Federal Aviation Administration (2015b) Unmanned aircraft systems (UAS) frequently asked questions. Washington, DC. Retrieved 22 Feb 2015, from https://www.faa.gov/uas/faq/#qn1

Flener C, Vaaja M, Jaakkola A, Krooks A, Kaartinen H, Kukko A, Kasvi E, Hyyppa H, Hyyppa J, Alho P (2013) Seamless mapping of river channels at high resolution using mobile LiDAR and UAV-photography. Remote Sens 5(12):6382–6407. doi:10.3390/rs5126382

Giles C (2014) Drones could predict region's lake effect snow. Great Lakes Echo. Retrieved from http://greatlakesecho.org/2014/10/29/drones-could-predict-regions-lake-effect-snow/

Gruen A (2013) Next generation smart cities-the role of geomatics. Singapore-ETH Centre, Future Cities Laboratory, Singapore

Haarbrink RB, Eisenbeiss H (2008) Accurate DSM production from unmanned helicopter systems. Int J Photogramm Remote Sens Spat Inf Syst XXXVII(Part B1):1259–1264

Jacques C (2014) Led by agriculture, market for commercial drones will reach $1.7 billion in 2025 [Press release]. Retrieved from http://www.luxresearchinc.com/news-and-events/press-releases/read/led-agriculture-market-commercial-drones-will-reach-17-billion

Jansen B (2015) FAA unveils drone rules; Obama orders policy for agencies. Retrieved from http://www.usatoday.com/story/news/2015/02/15/faa-drone-rule/23440469/?utm_source=dlvr.it&utm_medium=twitter&dlvrit=206567

Keane JF, Carr SS (2013) A brief history of early unmanned aircraft. Johns Hopkins APL Tech Dig 22(3):558–571

Linkhorn T (2015) Money-making potential aided by possible regulation changes for drone users. Toledo Blade. Retrieved from http://www.toledoblade.com/Economy/2015/02/22/Sky-s-no-limit-for-drone-manufacturers.html

Mancini F, Dubbini M, Gattelli M, Stecchi F, Fabbri S, Gabbianelli G (2013) Using unmanned aerial vehicles (UAV) for high-resolution reconstruction of topography: the structure from motion approach on coastal environments. Remote Sens 5(12):6880–6898. doi:10.3390/rs5126880

Martinez-De Dios JR, Ollero A (2006) Automatic detection of windows thermal heat losses in buildings using UAVs. Paper presented at the World Automation Congress, Budapest

Mulero-Pázmány M, Negro JJ, Ferrer M (2014) A low cost way for assessing bird risk hazards in power lines: fixed-wing small unmanned aircraft systems. J Unmanned Veh Syst 02(01):5–15. doi:10.1139/juvs-2013-0012

O'Callaghan J (2014) Planner tech: 5 digital city trends for 2015. Newplanner, December 2014, p 16

Ro K, Oh J-S, Dong L (2007) Lessons learned: application of small UAV for urban highway traffic monitoring. 45th AIAA aerospace sciences meeting and exhibit, 8–11 January 2007, Reno, Nevada. American Institute of Aeronautics and Astronautics, Reston

Sarda-Palomera F, Bota G, Vinolo C, Pallares O, Sazatornil V, Brotons L, Gomarez S, Sarda F (2012) Fine-scale bird monitoring from light unmanned aircraft systems. Int J Avian Sci 154:177–183

Schoellig A (2014) The role of unmanned aerial vehicles in future urban environments. Institute for Aerospace Studies, University of Toronto, Toronto

Tahar KN (2012) Aerial terrain mapping using unmanned aerial vehicle approach. International archives of the photogrammetry, remote sensing and spatial information sciences, volume XXX-IX-B7, 2012. XXII ISPRS Congress, Melbourne, 25 August – 1 September 2012, pp 493–498

Yuhas C (2006) Earth observations and the role of UAVs: volume 2 appendices. Civil UAV Assessment Team, NASA

Whitehead K, Hugenholtz CH, Myshak S, Brown O, LeClair A, Tamminga A, Barchyn TE, Moorman B, Eaton B (2014) Remote sensing of the environment with small unmanned aircraft systems (UASs), part 2: scientific and commercial applications. J Unmanned Veh Syst 02(03):86–102. doi:10.1139/juvs-2014-0007

Woody T (2014) Drones are becoming energy's new roustabouts. The New York Times, 22 April 2014

Xie T, Liu R, Hai RT, Hu QH, Lu Q (2013) UAV platform based atmospheric environmental emergency monitoring system design. J Appl Sci 13(8):1289–1296

Chapter 15
Urban Greening as a Social Movement

Sara S. Metcalf, Erika S. Svendsen, LaDona Knigge, Hua Wang, Harvey D. Palmer, and Mary E. Northridge

Abstract The premise of this chapter is that human engagement of city dwellers in acts of urban greening, or environmental stewardship, collectively constitutes a social movement for greater access to the natural environment. The power of urban greening as a social movement is particularly salient for disadvantaged communities in promoting greater resilience, health, and well-being. Grounded in the geographies of urban greening associated with Buffalo and New York City, we consider how some civic environmental stewardship programs positively impact youth. More specifically, we contend that such programs provide vulnerable populations such as adolescents with opportunities for social interaction that include intergenerational bridges and mentoring relationships that confer the psychological resilience needed to sustain such local activism. We develop a systems perspective illustrating how participants in successful civic environmental stewardship programs develop an enhanced sense of control and belonging to a community. The connections that

S.S. Metcalf (✉)
Department of Geography, The State University of New York at Buffalo, Buffalo, NY, USA
e-mail: smetcalf@buffalo.edu

E.S. Svendsen
U.S. Forest Service Northern Research Station, New York, NY, USA
e-mail: esvendsen@fs.fed.us

L. Knigge
Department of Geography and Planning, California State University at Chico, Chico, CA, USA
e-mail: lknigge@csuchico.edu

H. Wang
Department of Communication, The State University of New York at Buffalo, Buffalo, NY, USA
e-mail: hwang23@buffalo.edu

H.D. Palmer
Department of Political Science, The State University of New York at Buffalo, Buffalo, NY, USA
e-mail: hpalmer@buffalo.edu

M.E. Northridge
Department of Epidemiology and Health Promotion, New York University College of Dentistry and Global Institute of Public Health, New York, NY, USA
e-mail: men6@nyu.edu

J.D. Gatrell et al. (eds.), *Urban Sustainability: Policy and Praxis*, Geotechnologies and the Environment 14, DOI 10.1007/978-3-319-26218-5_15

individual residents feel to their surrounding community manifest in a heightened concern for others and increased activism promoting public access to green space.

Keywords Local environmental stewardship • Urban green space • Youth engagement • Community health • Systems science

15.1 Introduction

Why don't we see more green space in urban areas? Why doesn't public support arise naturally for local government to prioritize greater access to green space? Given its widely accepted benefits, green space is arguably a public good that is undersupplied. In other words, we cannot expect political coalitions to emerge spontaneously to pressure for local government to change status quo policies on land use, and thus activism for urban greening is not only necessary but crucial.

This chapter focuses on the role of community greening initiatives in fostering social movements comprised of activism for access to green space. We illustrate facets of urban greening in different ways: using geographic maps of Buffalo, NY and New York City, developing a causal map of youth engagement in greening from a systems perspective, and proposing a research design for further inquiry. We contrast urban greening efforts in different geographic contexts to identify relationships between factors of youth engagement with environmental stewardship programs. Through the use of a causal map illustrating complex feedback mechanisms, this study develops a systems perspective to explore the urban geographies associated with environmental stewardship programs that engage youth. Using a systems science approach, we articulate salient pathways for youth engaged in local environmental stewardship programs. These pathways produce complex feedback effects that amplify engagement with urban greening programs. When sustained through civic environmental stewardship, green space affords communities with opportunities for their residents to express, support, heal, and inspire one another (Svendsen et al. 2014).

A core experience of those involved in the social movements underlying civic environmental stewardship is that of the expansion of their individual perspective to become part of a shared point of view. Once activated, the interpersonal relationships arising from successful stewardship programs aggregate across scales to characterize facets of urban greening as a larger social movement. Here we wonder whether the versatility of this sort of social movement lies in the agility of its opinion leaders, acting at the nodes of the larger greening network, to effectively leverage the more fragile relationships that connect local actors strewn across a wide range of urban environmental stewardship organizations (e.g., community gardens) and other human catalysts of urban greening (e.g., the Massachusetts Avenue Project[1] in Buffalo).

[1] The website for the Massachusetts Avenue Project (MAP) is: http://www.mass-ave.org. This non-profit organization engages youth in urban greening and has been studied as part of Buffalo's local food movement (Metcalf and Widener 2011; Metcalf 2012).

Green space is important and beneficial for exposing community members to the natural environment (Marcus and Sachs 2014). Time spent outdoors provides opportunities to explore the many intersections of grey and green infrastructure and to become aware of the local histories that are embedded into the ordinary geographies of everyday experience (Stilgoe 1998). In this chapter we examine the particular benefits of youth participation in community greening initiatives for health and well-being. Indeed, the concept of health is a confluence of physical, mental, social, and spiritual well-being. Factors at multiple scales contribute to both individual health outcomes and societal inequalities (Northridge et al. 2003).

Western society tends to emphasize the utilitarian value of parks and other kinds of green space, assessing their value in terms of ecosystem services, recreational use, or real estate value. Community gardens, for example, may be seen as having an agricultural or aesthetic value. Yet, some local governments may also take into account the opportunity cost of using the community garden land as residential or commercial space, which is realized as lost tax revenue and potential jobs. This utilitarian perspective helps to account for the rise of urban farms in "hollowed out" parts of old industrial cities such as Buffalo where the tax base has been shrinking, so the opportunity cost is low for the local government.

Green space can be the site of shared understanding, but the ownership of such space may be contentious. It is contentious both instrumentally and spiritually. Instrumentally, responsibility for green space is shared and hence when the sense of community is lacking, neglect often follows. Spiritually, parks as public spaces often host gatherings for political reform and social protest, as illustrated in recent years by the Occupy Wall Street Movement and the conflict over Gezi Parke in Istanbul.

When aligned with purposeful human activity, a public place can be transformed from ordinary to sacred (Northridge and Mark 2013). The land's value then goes beyond the utilitarian, since the the rating of its use brings psychological benefits that cannot be easily replaced. Research has explored the spatial contours of significance that arise from one's proximity to, and the nature of, human loss, as in living memorials created in remembrance of the tragic events of September 11, 2001 (Tidball et al. 2010; Svendsen et al. 2014).

Local environmental stewardship signifies a special kind of care in the accumulated acts of giving back that serve to interconnect humans with our grey (i.e., built environment) and green (i.e., natural environment) urban infrastructures. Human beings thus function as "inter-actors" embedded in a social, ecological urban system. The cultivation of green space compels human attention and care, invites access, and encourages social interaction. This interconnectedness catalyzes dynamics that fuel activism for access to green space.

Green space promotes many positive health outcomes through exposure to nature as well as opportunities for social connection. The restorative and therapeutic capacity of green space has been well documented (Marcus and Sachs 2014). In particular, the mental health benefits of green space have been recognized and found to vary with the life course (Astell-Burt and Feng 2015; Astell-Burt et al. 2014). As explored in this study, human experiences with stewardship of the natural environment serve to instill a sense of belonging to one's community.

Resilience requires anticipating change in the habitability of places, such as coastal zones. It means being able to adapt, learning to leave a living legacy for humankind. Civic environmental stewardship, in this sense, is a salvation strategy. Recognizing nature as an integral part of social systems, we locate the potential for resilience in social relationships as well as in the physical design and form of urban environments (Svendsen et al. 2014).

This chapter is oriented toward the personal relationships activating social movements promoting greener urban geographies and ways that people can positively impact the production of ecosystem services. In the least green places there are sometimes higher levels of stewardship than expected (Tidball and Krasny 2014). This tendency suggests a potential feedback mechanism by which a dearth of community resources produces a counteracting impulse to heal and help one another. If greening happens more where it is most needed, as in red zones (Tidball and Krasny 2014), then the need (the gap) would incrementally decrease as greening efforts become more prevalent, iteratively closing the gap. Consistent with this logic, we use systems science to explore ripple effects and identify leverage points for urban greening activities such as community gardening (Svendsen et al. 2012).

15.2 Geographic Contexts

When developing a more comprehensive theoretical model, it is often useful to first explore a few cases that are reliable microcosms of the system being studied but which differ in their geographic context. Given our focus on urban greening, we compare two geographic contexts that highlight structure differences among urban areas more generally. Specifically, we consider how urban greening constitutes a social movement in the geographic contexts of Buffalo and New York City in the state of New York. While Buffalo and New York are the two largest cities in the state of New York, they offer a stark contrast in scale as they straddle the urban hierarchy, with New York City's population of 8.4 million residents dwarfing the city of Buffalo's population of approximately 260,000 residents. Despite their marked differences in population size and density, both cities have experienced urban greening as an emergent social movement. Here we discuss some particulars of these geographies of urban greening. These cases provide a foundation for the systems perspective of youth engagement in greening programs that we develop using a causal map in the subsequent section.

15.2.1 Community Gardens in Buffalo

Green space confers diverse utilities, evidenced by the symbols of abundance such as flowers and fruit that are associated with green space. Although social meanings evolve over time, a common thread associated with gardens and other kinds of green space is that of care, or concern, for one another (Knigge 2009). Community gardens offer tangible nutritional benefits to residents in the form of fresh fruits and vegetables, but also confer intangible benefits like the aesthetic appreciation and sense of sacredness that many feel in the presence of nature (Svendsen 2011; Svendsen et al. 2012). This chapter builds upon earlier systems explorations of the social movements promoting local food and urban agriculture in Buffalo, NY (Metcalf and Widener 2011; Metcalf 2012). This iterative examination aligns with the approach taken by Knigge and Cope (2009) to study green space.

Here, our consideration of the utility afforded by green space begins with a 2004 inventory of community gardens in Buffalo, NY that was mapped as part of a study that employed iterative qualitative GIScience for grounded visualization of urban greening initiatives (Knigge 2009; Knigge and Cope 2009). The map in Fig. 15.1 illustrates the locations of community gardens in Buffalo, NY as characterized by purpose: ornamental, food, or both. While many of the gardens mapped in Fig. 15.1 have an ornamental purpose, some also provide food. This visualization highlights the different types of gardens that may be cultivated for and by the community. The varied types of use and lack of a clear geographic pattern in that use reflects the community impetus and local context that shapes use decisions.

As the presence of community gardens and green space is depicted in Fig. 15.1 alongside median income according to the 2000 U.S. Census, we observe a relative dearth of both in the poorest areas. This points to the importance of leveraging urban greening as a social movement to better serve disadvantaged communities. Additionally, the similar lack of community gardens in the most affluent areas suggests that urban greening arises more naturally as a social movement in middle-class communities.

15.2.2 Environmental Stewardship in New York City

Shifting scale to New York City enables us to consider different kinds of insight about urban greening activities. In this section we draw upon the rich data collected by the Stewardship Monitoring and Assessment Project (STEW-MAP) for New York City.[2] Through organizational surveys and interviews, STEW-MAP explores the interactions among people and groups that conserve, manage, monitor, advocate for, and educate the public about their local environments. STEW-MAP includes voices from citizen groups in New York City such as informal block associations and tree planting groups as well as formal nonprofit organizations. Studies leveraging NYC STEW-MAP observations highlight the importance of stewardship as a form of governance and the growing professionalization of these civic groups (Svendsen and Campbell 2008; Connolly et al. 2014). STEW-MAP research also reveals how stewardship storylines influence design and practice (Svendsen 2013). These STEW-MAP studies characterize the structure and function of urban environmental

[2] The STEW-MAP NYC homepage: http://www.stewmap.net/stew-map-cities/new-york-city/

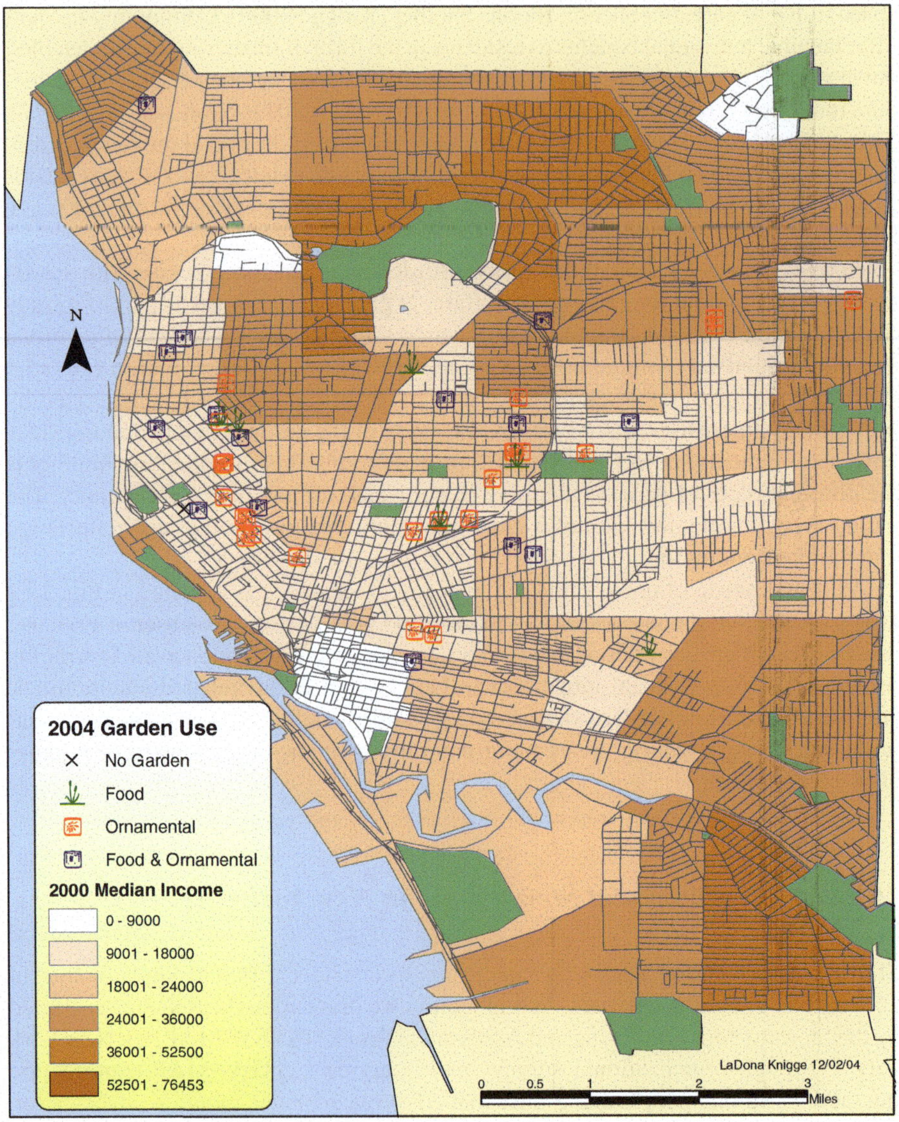

Fig. 15.1 Community gardens in Buffalo, NY

stewardship and the organization of stewardship networks in the city (Svendsen and Campbell 2008). Network analysis identifies the connections between civic environmental groups and key stewardship nodes that work with many different partners (Connolly et al. 2013, 2014). Through its rigor and relational emphasis, the STEW-MAP project effectively adds a social layer of information about green urban geographies in New York City.

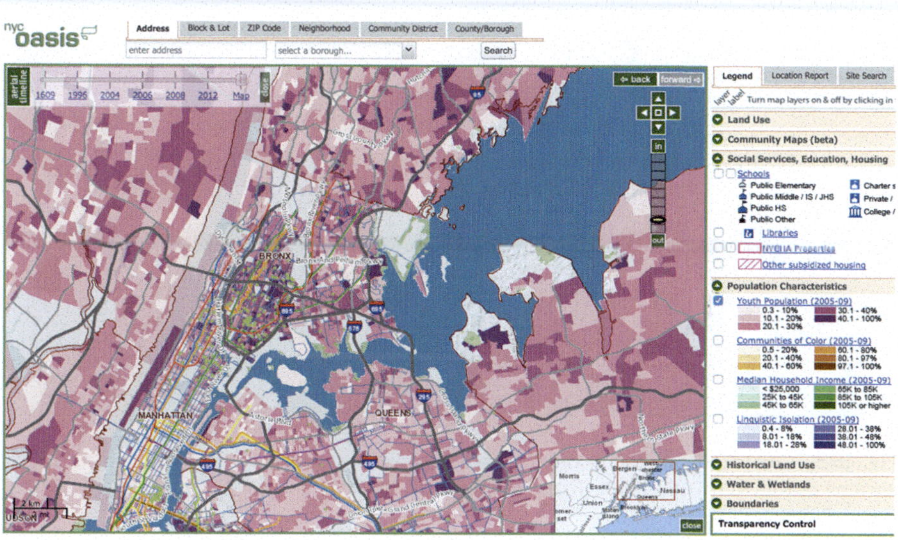

Fig. 15.2 Interactive GIS featuring layers for youth population density and green space

An accessible and iterative investigation of stewardship in New York City is enabled by STEW-MAP's online GIS database, which displays geographic data associated with many of these stewardship groups alongside other data layers. Figure 15.2 is an example of the custom maps that can be constructed using the online GIS provided via the STEW-MAP project, linking information about stewardship areas and organizations with other geospatial data.[3] The map in Fig. 15.2 includes a layer for the prevalence of youth in the population of each census tract so as to facilitate exploration of where opportunities exist for urban youth to be exposed to green space and to engage in acts of civic environmental stewardship. The many health benefits of green space for youth are described by Louv (2008). These benefits warrant the development of more stewardship programs in underserved communities that connect youth to their local environments so as to promote health equity.

In addition to the demographic layers that can be illustrated, as in Fig. 15.2, the STEW-MAP project also enables mapping of stewardship "turf," or areas of operation for civic environmental stewardship organizations. The turf maps are based upon answers to a question in which survey respondents were asked to describe in detail the boundaries of where their group works. For example, Fig. 15.3 contrasts turf for stewardship programs that primarily focus on education (at left) and youth (at right).[4]

[3] Figure 15.2 was created using the publicly available online database from STEW-MAP NYC: http://www.urbanresearchmaps.org/oasis/map.aspx

[4] The maps in Fig. 15.3 were created using the publicly available online database from STEW-MAP NYC: http://www.oasisnyc.net/stewardship/stewardshipsearch.aspx

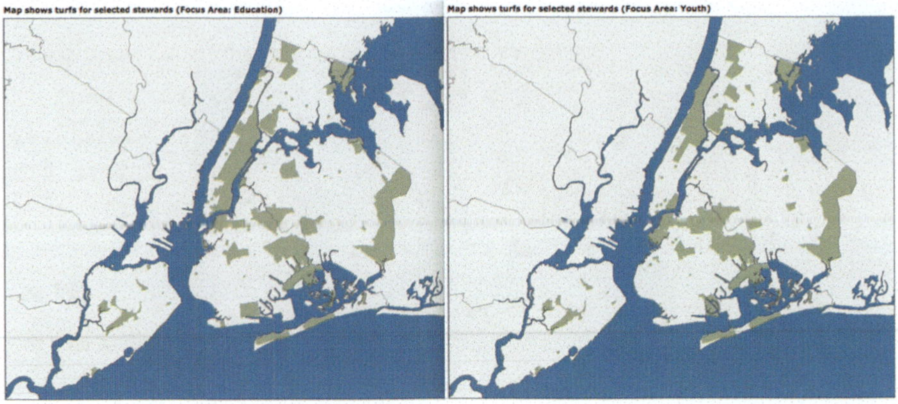

Fig. 15.3 Turf maps for civic environmental stewardship organizations in New York, NY that mention education (*L*) and youth (*R*) as primary focus areas

The maps in Fig. 15.3 illustrate some areas of overlap in the spatial distribution of the turf associated with stewardship programs that indicate education and youth as primary areas of focus. Online maps like these may be used by the public to examine the spatial influence of organizations that engage with youth in their greening initiatives. For instance, while it is natural to expect high clustering of the activities of education and youth organizations given their similar focus, Fig. 15.3 identifies significant areas of turf that are geographically distinct. This pattern is consistent with theories of social movements in that these organizations often seek to establish reliable bases of support within the community by leveraging existing social institutions and other connections to local social networks. These maps demonstrate the benefit from shifting perspectives by which one could take an iterative triangulation approach to contrasting landscapes of urban greening.

15.3 Human Experiences with Urban Green Space

Human beings are increasingly facing the consequences of global climate change in terms of extreme events experienced as sudden environmental shocks to the system. Our society continues to experience more uncertainty. We have undergone a shift in awareness of our societal vulnerabilities and are learning how to work more effectively with our grey, green, and human resources. To address the sustainability challenges of our time, we must understand the implications of a growing diversity of populations needing access to green space. For example, although there have been growing investments in New York City's green space, many of these are motivated by preservation of capital property, so that there are stark inequities in the conditions of parks across the city. Here we may ask: Why aren't we realizing the environmental and community health outcomes we want? What would the city look like if we managed green space to produce different outcomes?

Observational studies of green space usage reveal diverse park users who vary by time of day. In making such observations, we might wonder: What is expressed when a person is observed to be sitting on a city park bench? An older person may be communing with a younger self, reconnecting with memories of earlier stewardship activities. Honoring, being, reflecting: these subtle sacred practices lack a recognizable language. How do we adequately account for the value of simply relaxing and enjoying being at the park?

What do we know about sacred space? Speaking of the sacred transforms our conception of well-being into something we haven't been able to easily talk about. How is the sacred expressed? There is a layered, hidden, everyday meaning expressed by human activity in green space. Acknowledging the sacred goes beyond the utilitarian view to say that spiritual human needs are also fulfilled by the park and the garden. These experiences feed back to nourish a stronger self. This view positions people themselves as part of the ecosystem services they produce.

15.4 Modeling Systems of Boxes, Bodies, and Botany

Whether we visualize our cities as grey, green, or tinted with other hues, we sometimes share, via our unseen interconnections, a sacred feeling. Using the tools of systems science, we might trace a sacred spiral as an ongoing cycle involving ripple effects from the transformation of ordinary to sacred in human perceptions of place. A person might trigger a reinforcing feedback loop promoting a clearer conscience and sharper perception of oneself in one's surroundings over time, facilitating the adoption of more sustainable behavior. This conception encompasses diverse but interdependent elements of the urban setting, from intangibles such as human relationships and mental models to tangible gardens, groves, graves, roads, rivers, skyscrapers, vehicles, and the like. With the aid of systems modeling, we might simulate our cities using discrete objects for boxes (e.g., "grey" buildings and the like), bodies (e.g., human beings and other animals), and botany (e.g., "green" plants in gardens, parks, and other urban spaces).

System dynamics is an established methodology for modeling complex systems involving specification and simulation of feedback mechanisms comprised of causal linkages and time delays (Sterman 2000). In this modeling tradition, boxes are used to represent continuously changing stocks, or accumulations, of quantities that characterize the state of the system. A environmental stock relevant to urban greening would be the local biodiversity of the urban ecosystem as influenced by the flow of stewardship activity. Intangible stocks include human memory, considered at both the collective and the individual scale. Human experience, as in exposure to green space or to pollution, also constitutes a stock that accumulates over time. Social capital could also comprise a stock of accumulated interpersonal interactions.

In contrast to stocks, flows indicate rates of change over time that add to or deplete the stocks to which they are connected. Green space, for example, may be conceptualized as a stock, while cultivation (greening) is a flow into the stock. As depicted in Fig. 15.4, this stock of green space may be diminished by urban devel-

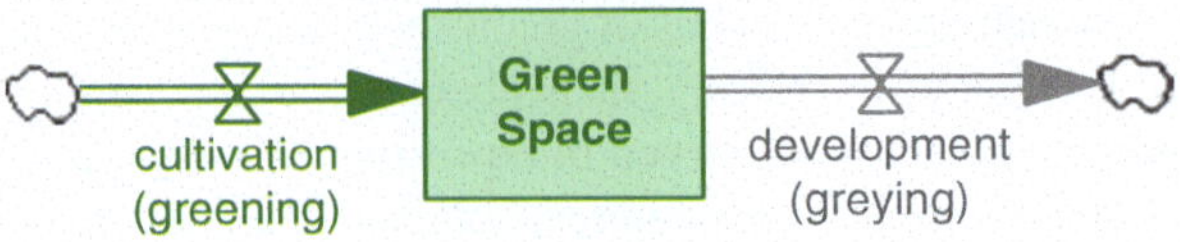

Fig. 15.4 Green space conceptualized as a stock changing via flows

opment (greying). This representation is a simplified view that does not account for the multiple ways in which grey and green infrastructures may be integrated (as explored in Svendsen 2011).

Human activity is incurred over time as a flow of energy invested in our urban infrastructures. If human resources are considered as a kind of distributed energy, it follows that a social movement carries a flow of energy distributed among our human resources. Thinking about flows as rates of change helps to differentiate between rapid/acute (shock) or slow/chronic (abandonment) system disturbances. In coastal regions, for instance, we would be interested in the effects of rapid shocks such as storm surges as well as more chronic disturbances such as sea level rise.

15.5 A Systems Perspective on Youth Engaged in Greening

An important component of this systems science approach is to make visible the unseen. In drawing out these invisible relationships, we use the language of systems thinking to express qualitative insights. As an integral part of public engagement with local environmental stewardship, we emphasize here the significance of youth who engage in greening activities. Adolescence is a particularly vulnerable stage in the life course (Steinberg 2014). Effective interventions with adolescents therefore have the potential to change the long-term life trajectories of youth participants. The systems perspective developed in this section connects outer activism with the inner psychological resilience of youth via multiple pathways.

Figure 15.5 illustrates such a causal map of youth engagement in greening initiatives.

This causal map visually expresses a dynamic hypothesis as a set of theoretical expectations about relationships among factors that motivate youth engagement with community greening initiatives. While the map is an integrated whole, it is "built" from deeper, theoretical reflection (informed by extant research) about the causal process and the nature of feedback mechanisms involving local social networks, institutions, and organizations.

Development of a causal map such as that depicted in Fig. 15.5 is consistent with the practice of system dynamics and follows the principles of systems thinking (Meadows 2009). A feedback loop is reinforcing if it has an even number of negative links, and balancing if it has an odd number of negative links. For the causal map in Fig. 15.5, all links are positive, so all feedback loops have zero (0) negative links. Because zero is an even number, all loops are reinforcing. Therefore, although positive and negative causal links are often depicted with positive (+) and negative (−)

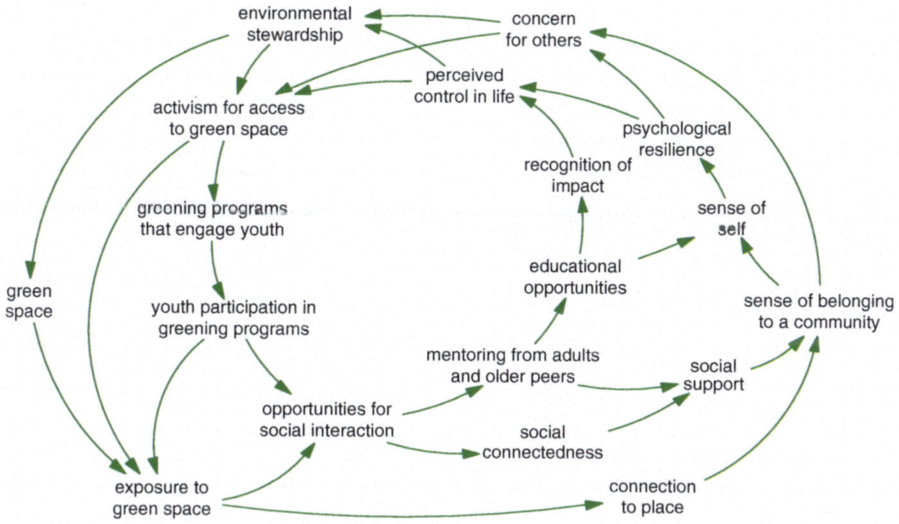

Fig. 15.5 Reinforcing psychological resilience in youth through environmental stewardship

polarity signs on the arrows, for simplicity no polarity signs are shown in Fig. 15.5 since all links are positive.

The causal relationships drawn in Fig. 15.5 produce a set of reinforcing feedback loops. In describing these mechanisms for the case of youth engagement in greening, we trace our logic in terms of increasing growth, but note that the reinforcing tendency also applies to decreases in the corresponding elements. In other words, the reinforcing feedback effect amplifies the change in the same direction, whether increasing or decreasing. Reinforcing feedback loops produce opposite behavior from that of balancing feedback mechanisms in which changes to the system in one direction are counteracted by their effects in the opposite direction. In other words, the reinforcing loops cause effects to be multiplicative in nature where change in one part of the system is strengthened by triggering subsequent change elsewhere in the system.

Beginning at the top of Fig. 15.5 with *environmental stewardship* and tracing the outermost loop counterclockwise, we follow the logic that an increased abundance of *green space* leads to more human *exposure to green space*, thereby enhancing one's *connection to place* and bringing about a *sense of belonging to a community*. Such a sense of belonging links the individual to their community, increasing *concern for others*, which in turn increases likelihood of more *environmental stewardship* and also *activism for access to green space*.

From this effect and the connection between *environmental stewardship* and *activism for access to green space*, another set of reinforcing loops is then activated. Following the innermost loop, with an increase in public *activism for access to green space*, we expect to see more urban greening programs, and specifically *greening programs that engage youth*. A greater abundance of urban greening programs leads to more *youth participation in greening programs*. In addition to the environmental benefits of civic stewardship, then, we see that youth participation leads to increased *opportunities for social interaction*. Indeed, effective environmental stewardship programs have learned that these opportunities include *mentoring from adults and older peers*. Over time, such mentoring improves *educational opportunities* for disadvantaged youth. Consistent with the differences in "turf" illustrated in Fig. 15.3, we expect youth- and education-focused greening programs to serve overlapping but distinct populations due to the effects in this reinforcing loop.

The broadened horizons of participants in these educational programs help to bring about the *recognition of impact* that one's environmental stewardship activities can have on overall community health and well-being. Such a personal recognition would enhance one's *perceived control in life* and thereby feed back to increased *activism for access to green space*. Tracing this feedback loop highlights how we believe activism for access to green space will have ripple effects in a reinforcing manner that eventually lead to more activism.

Additional feedback effects are created as elements of the innermost loop connect with those of the outermost loop. For example, a cascading effect from increased activism and youth participation (along with the extent of green space itself, as noted above) is increased *exposure to green space*. Another loop may be traced counterclockwise from increased *youth participation in greening program* to a strengthened *connection to place*.

Following again from increased youth participation in greening programs, for the third loop we look to the *opportunities for social interaction* that occur in the context of a community greening initiative. All else being equal, this increased opportunity leads to greater *social connectedness*, *peer acceptance*, and *sense of belonging*. These social opportunities can also lead to *mentoring from adults and older peers*, which in turn benefits *social support* as well as *educational opportunities* (as noted above). *Social support* leads to a greater sense of belonging to a community. This sense of belonging, along with expanded *educational opportunities* from effective mentoring ultimately enhances one's *sense of self*.

A critical effect from *sense of self*, then, is to strengthen *psychological resilience*, which feeds back to both *perceived control in life* and *concern for others*, completing multiple reinforcing feedback loops. Through these feedback loops, *sense of self* plays a central role in improving the mental health of participants in urban greening programs.

Figure 15.5 positions *psychological resilience* as a key component of systemic health and well-being. Specific mental health outcomes challenging psychological resilience include depression and anxiety. Numerous other positive effects of green space on health and well-being are known to exist, such as health benefits from physical activity associated with stewardship and recreation, or nutrition benefits from improvements to healthy food access through community gardens. These alternate pathways can be thought of as additional reinforcing links between *exposure to green space* and *psychological resilience*.

In short, the causal map specifies how urban greening produces health benefits such as psychological resilience and social well-being. Urban greening programs also impact the educational and life goals of young people. Stewardship networks form generational bridges such as adults and older peers mentoring vulnerable youth.

In the following section we outline a multi-method research design that builds upon the conceptual framework presented above to further study success factors of youth engagement in stewardship programs.

15.6 Looking Ahead: A Design for Research on Greening and Well-Being

Our exposition of these geographic and causal maps provides a foundation for future research focused on modeling youth engagement with greening programs in disadvantaged communities. The important work of urban greening needs a health equity lens. While the link between greening and health has been well studied (see, e.g., Astell-Burt et al. 2014), this line of research is unique and significant in terms of its emphasis on promoting health equity and opportunities for young people who live in disadvantaged communities. We must ensure that groups who would most benefit from salutogenic design are actually part of the decision-making process.

The complexity of achieving public health equity is particularly salient in communities with a diverse identities, needs, and aspirations. Greening is high leverage in promoting health not just because of increased exposure to urban nature but also because of the experience of engaging with a distinctive social movement. Following from the causal map depicted in Fig. 15.5, this study aims to conduct additional research to understand youth-oriented environmental stewardship programs that are successful. Specifically, we seek to characterize what the causal mechanisms are in an understudied population of adolescents who don't have a lot of resources.

A benefit of qualitative research, and of systems science, is that insights emerge in the course of conducting the research. For this reason, community-based system dynamics modeling involves iteration and inclusion of needed voices (Hovmand 2014). Yet we may ask: How do we integrate rich qualitative information into urban ecosystem management regimes? Despite its recognized importance for understanding urban greening, qualitative research is often difficult for natural resource managers of urban ecosystems to act upon. Nevertheless, qualitative data such as stories, interviews, social relations, and photographs are invaluable for the characterization of site-specific geographies of urban green space. These observations are complementary to quantitative measures and help to inform causal pathways and directions of system change.

This research will integrate participatory methods of systems science with a positive deviance approach to community health. Positive deviance is an asset-based approach to health behavior and social change, centered on collective discoveries of often microscale, invisible, yet highly effective solutions that already exist within the community without extra resources (Zeitlin 1991; Sternin and Choo 2000; Singhal et al. 2010). This research will explore behaviors of stewardship organizations that exhibit positive deviance.

This inquiry into the effectiveness of urban greening programs that engage with youth in disadvantaged communities will leverage the STEW-MAP social ecosystem database of stewardship organizations in New York City introduced earlier. Our first step will be to follow up with the originally surveyed participants to identify stewardship organizations exhibiting positive deviance that maintain successful community partnerships and engage with youth. From a set of organizations that exhibit positive deviance, participatory workshops will be conducted to discover and visually map out what organizational and individual practices can, against all odds, lead to youth engagement in greening programs. Furthermore, these community-based participatory workshops will explore how youth engaged in greening activities abet their mental health and social connectedness.

Building upon the conceptual framework outlined in Fig. 15.5, these more complete workshops and youth survey data will inform the development of a model of youth engagement in greening programs. In particular, this research will consider the attitudes, experiences, and mental health outcomes – specifically depression and anxiety – among youth participating in urban greening programs. These outcomes are amenable to support from such initiatives. Mental health has other connections to physical activity, use of coping mechanisms, and formation of social ties, which may also be strengthened (improved) by exposure to greening programs.

This line of research leverages benefits of hybridity, using multiple methods to gain a more expansive view of the system. Specifically, we aim to combine the rigor of positive deviance to select environmental stewardship organizations applying participatory approaches to translate qualitative insights into a model. This research will inform the design of a multiscaled model to illustrate the structure of the system and explore the social and economic dynamics underlying the availability and utilization of urban green space. Both qualitative and quantitative GIS data can inform or be directly integrated into a dynamic model. Our iterative multi-method approach facilitates grounding abstract models with empirical observations and offering the potential to visualize changes in behavior over time.

15.7 Conclusion

The provision of green urban space, particularly in socioeconomically disadvantaged communities, creates opportunities for social behaviors that promote well-being and health equity. For many people green space even reflects a sense of the sacred, an intangible connectedness, a kind of "mad love," that transcends the ordinary. These places are therefore integral not only to the daily lives of community members, but also to the collective spirit of humanity. Communities seeking to promote resilient green infrastructure would be well-served to nurture the social meaning in these shared places.

This study points to the importance of new research to identify and model effective strategies for engaging youth in greening programs in disadvantaged communities. Such research would examine which environmental stewardship organizations particularly excel at engaging disadvantaged youth in greening projects without extra resources. In turn, these organizations will provide unique insight into efficient and effective strategies to improve the mental health of urban youth through involvement in greening programs.

In articulating urban greening as a social movement, we envision the cultivation of neighborhood resilience networks in which people are active agents. In this era of technological advancement, social inequality, and environmental crises, we have a responsibility to act with awareness of our human contributions to the global ecosystem. We find evidence of the social dimension of sustainability in the interrelationships that underlie successful environmental stewardship programs and their benefits for health and well-being among all members of the community. This social connectedness is a source of hope for increased activism over time for greater access to green space in urban areas due to the positive mental health benefits.

Acknowledgments Work on this chapter was supported in part by the USDA Forest Service (Integrating Grey and Green Infrastructure to Improve Health and Well-Being for Urban Populations, award 12-JV-11242309-095) and the National Institute of Dental and Craniofacial Research and the Office of Behavioral and Social Sciences Research of the US National Institutes of Health (Integrating Social and Systems Science Approaches to Promote Oral Health Equity, award R01DE023072).

References

Astell-Burt T, Feng X (2015) Investigating 'place effects' on mental health: implications for population-based studies in psychiatry. Epidemiol Psychiatr Sci 24:27–37

Astell-Burt T, Mitchell R, Hartig T (2014) The association between green space and mental health varies across the lifecourse. A longitudinal study. J Epidemiol Community Health 68:578–583

Connolly JJT, Svendsen ES, Fisher DR, Campbell LK (2013) Organizing urban ecosystem services through environmental stewardship governance in New York City. Landsc Urban Plan 109:76–84

Connolly JJT, Svendsen ES, Fisher DR, Campbell LK (2014) Networked governance and the management of ecosystem services: the case of urban environmental stewardship in New York City. Ecosyst Serv 10:187–194

Hovmand P (2014) Community-based system dynamics. Springer, New York

Knigge L (2009) Intersections between public and private: community gardens, community service and geographies of care in the US City of Buffalo, NY. Geogr Helv Swiss J Geogr 64:45–52

Knigge L, Cope M (2009) Grounded visualization and scale: a recursive analysis of community spaces. In: Cope M, Elwood S (eds) Qualitative GIS: a mixed methods approach. Sage, Thousand Oaks, pp 92–114

Louv R (2008) Last child in the woods: saving our children from nature-deficit disorder. Algonquin Books, Chapel Hill

Marcus CC, Sachs NA (2014) Therapeutic landscapes: an evidence-based approach to designing healing gardens and restorative outdoor spaces. Wiley, Hoboken

Meadows DH (2009) Thinking in systems: a primer. Chelsea Green, White River Junction

Metcalf SS (2012) A systems modeling framework for the role of agriculture in a sustainable urban ecosystem. In: Reid N, Gatrell J, Ross P (eds) Local food systems in old industrial regions: concepts, spatial context, and local practices. Ashgate Press, Burlington, pp 6–28

Metcalf SS, Widener MJ (2011) Growing Buffalo's capacity for local food: a systems framework for sustainable agriculture. Appl Geogr 31(4):1242–1251

Northridge ME, Mark J (2013) Protecting public places. Am J Public Health 103(7):1157

Northridge ME, Sclar E, Biswas P (2003) Sorting out the connections between the built environment and health: a conceptual framework for navigating pathways and planning healthy cities. J Urban Health 80:556–568

Singhal A, Buscell P, Lindberg C (2010) Inviting everyone: healing healthcare through positive deviance. Plexus Press, Bordentown

Steinberg L (2014) Age of opportunity: lessons from the new science of adolescence. Houghton Mifflin Harcourt, New York

Sterman JD (2000) Business dynamics: systems thinking and modeling for a complex world. Mc-Graw Hill, Boston

Sternin J, Choo R (2000) The power of positive deviancy. Harv Bus Rev 78(1):14–15

Stilgoe JR (1998) Outside lies magic: regaining history and awareness in everyday places. Walker and Company, New York

Svendsen ES (2013) Storyline and design: how civic stewardship shapes urban design in New York City. In: Pickett STA et al (eds) Resilience in ecology and urban design: linking theory and practice for sustainable cities. Springer, Dordrecht, pp 269–287

Svendsen ES, Campbell LK (2008) Urban ecological stewardship: understanding the structure, function, and network of community-based urban land management. Cities Environ 1(1):1–31. http://digitalcommons.lmu.edu/cate/vol1/iss1/4/

Svendsen E (2011) Cultivating health and well-being through environmental stewardship. Am J Public Health 101(11):2008

Svendsen ES, Northridge ME, Metcalf SS (2012) Integrating grey and green infrastructure to improve the health and well-being of urban populations. Cities Environ 5(1). http://digitalcommons.lmu.edu/cate/vol5/iss1/3

Svendsen ES, Baine G, Northridge ME, Campbell LK, Metcalf SS (2014) Recognizing resilience. Am J Public Health 104(4):581–583

Tidball KG, Krasny ME (eds) (2014) Greening in the red zone: disaster, resilience, and community greening. Springer, New York

Tidball KG, Krasny ME, Svendsen E, Campbell L, Helphand K (2010) Stewardship, learning and memory in disaster resilience. Environ Educ Res 16(5–6):591–609

Zeitlin M (1991) Nutritional resilience in a hostile environment: positive deviance in child nutrition. Nutr Rev 49(9):259–268

Chapter 16
Decision Making and Sustainability in Built Environments

Michael J. Clay and Jenna N. Albers

Abstract Sustainability in the built environment is the result of choices make by those who plan, build, own, and use our urban spaces. This chapter seeks to detail how human choice shapes our built environment, as well as tools that can be used to create more sustainable urban futures. These tools include: (1) analytical forecasting models that are based upon urban economic theory and data provided by geotechnologies such as geographic information systems and remote sensing; (2) traditional zoning and other local land use policies that focus on protecting human health and property values that have resulted in an urban system that favors the private automobile as the dominate form of transportation and limits the viability of other forms of transportation such as mass transit, walking, and biking; (3) alternatives to traditional zoning such as mixed use and Form Based zoning practices that seek to improve quality of life and the sustainability of the built environment; and (4) the role that our individual and collective decisions play in the sustainability of our built environment. The concentration of people in urban areas places a strain on resources, both natural and built, that must be evaluated and balanced in order to maintain a sustainable habitat that not only provides all basic human necessities but respects the natural environment as well. Ultimately, the degree to which our built environment is sustainable has been and will continue to be the result of our individual and collective decisions that we make.

Keywords Urban planning • Urban sustainability • Built environment • Land use modeling

16.1 Introduction

Cities are the centers of innovation, trade, wealth, and prosperity in our modern world. Cities have greatly increased our quality of life and length of life through improved access to healthcare, sanitary drinking water, and the systematic removal

M.J. Clay, Ph.D. (✉) • J.N. Albers
Department of Geography, Brigham Young University, Provo, UT, USA
e-mail: clay@byu.edu; jen.nicole92@gmail.com

J.D. Gatrell et al. (eds.), *Urban Sustainability: Policy and Praxis*,
Geotechnologies and the Environment 14, DOI 10.1007/978-3-319-26218-5_16

of waste. Furthermore, cities have expanded our opportunities for education, employment, social interaction, and human networking. In short, cities are very, very good for humans (O'Sullivan 2006).

Cities are also good for the environment. By concentrating human populations in a relatively small geographic area, the impact of human settlement on habitats and migration corridors is minimized. The amount of infrastructure on a per person basis is also minimized, preserving valuable natural resources. While the impact on local habitat is devastating, the overall impact on a per person basis is diminished.

Our cities are not, however, without significant challenges. We consume massive amounts of non-renewable natural resources and create millions of tons of air, water, and ground pollution every year. Social inequality, poverty, and crime remain substantial issues in all urban areas. Our energy rich lifestyles and an over dependence on fossil fuels, particularly petroleum, as our dominate source of energy have created geo-political instability and the loss of human life. While our cities are better than the alternative (large populations spread out across a rural landscape) there remains much room for improvement.

It is within this context that this chapter will examine how better decision making can produce a more sustainable built environment to achieve more desirable urban futures. The goal of this chapter is twofold: first, to present the process of decision-making in the built environment as well as the institutions and individuals involved in this process; and second, to present a decision support tool that can assist in bringing about more sustainable urban futures.

16.2 The Role of Human Decision Making in the Built Environment

Human settlements are contrived. They do not happen in the absence of human involvement and human effort. Human settlements are the result of human decisions. Because human decisions determine the nature and extent of our built environment, what we see today as our "cities" or our built environment can accurately be attributed to decisions of the past. Where our cities will go in the future will be determined by the decisions we make today.

Decision-making in the built environment is guided by human laws and take place in a political environment. Understanding both is important. While human laws often determine the "rules" of the "game", they do not control the consequences of those decision. Fully understanding the consequences of our choices in the built environment remains the subject of much academic and practical research.

Human settlements represent a complex web of interrelated systems. These systems, such as transportation, economics, politics, social networks, land use, urban design, and the natural environment are each made up of smaller sub-systems that integrate with each other at various levels throughout the urban fabric. Often, policies seeking to address issues in one urban system unintentionally impact other urban systems.

Geotechnologies such as geographic information systems (GIS) and remote sensing are being employed in urban areas throughout the world to help monitor the spatial interactions of these urban systems and to assist in our understanding of the interrelationships among these systems. GIS, for example, is utilized as a decision support tool in US urban areas to help decision makers internalize the interrelationship been land use, transportation, the natural environment, and the economy.

16.3 Urban Sustainability

Urban sustainability can be difficult to define and even more difficult to achieve. The landmark 1987 report of the Brundtland Commission to the United Nations (World Commission on Environment and Development) defines sustainability as "development that meets the needs of the present without compromising the ability of future generations to meet their own needs." This is perhaps the most commonly used definition of sustainability. However, it leaves open two key issues: (1) any use of non-renewable natural resources (e.g. coal and petroleum) are resources not available to future generations—these are resources we are heavily dependent on to meet the needs of the current generation; and (2) a positivist approach (i.e. future technologies will enable future generations to meet their needs in ways we simply cannot understand or predict) enables us to continue consuming and polluting at current rates with little or no behavioral shifts.

The American Planning Association in 2000 produced the "Policy Guide on Planning for Sustainability" that catalogs many of the ways our built environment impacts sustainability. This guide also provides recommendations for improvement, acknowledging that human choices brought us to where we are in urban sustainability and that changing market preferences and government policy are required to achieve necessary improvements. This tact is the approach to sustainability used in this chapter.

16.4 Urban Decision Making

Humans make decisions that impact, shape, and construct our built environment. We do this in a variety of ways, settings, and roles (see Fig. 16.1). These include: users of the built environment; owners of the built environment; and as part of communities that have power over the built environment (i.e. government). Each of these is described below. The degree to which our built environment is sustainable is the culmination of the decisions we make in these various roles.

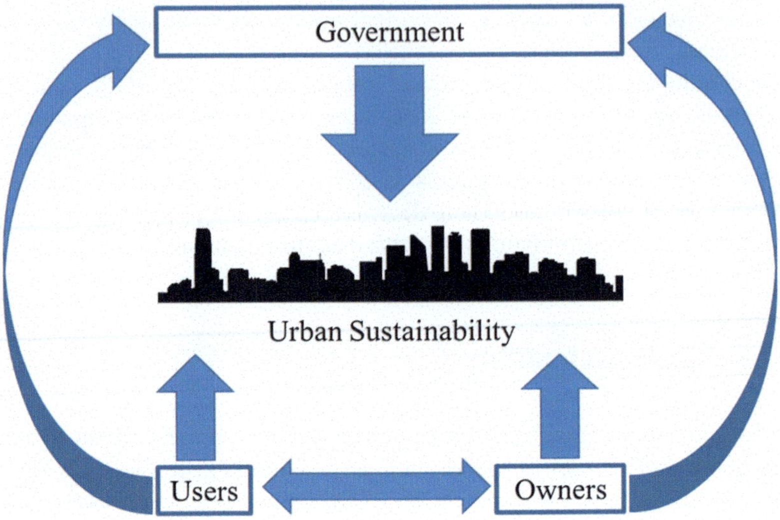

Fig. 16.1 Decision making in the built environment

16.4.1 Humans as Users and Occupiers of Space

Our most basic decision making role with regard to the build environment is as an occupier or user of land or space within a building. We live, eat, work, shop, visit doctors, dentists, or other professionals within the built environment every day. In this role, the patterns jointly developed by the user and provider of space are evident. The spatial arrangement of land use (residential, commercial, industrial, professional office, etc.), while not entirely market driven, does at least respond to market demand or in other words, the choices that we as users make.

How we choose to move between locations is also critical to urban sustainability. In the United States, this is primarily done with the private automobile. The consequences of this decision dominate our built form: wide streets with fast moving cars; narrow or non-existing sidewalks; huge parking lots—often larger than the store, bank, or office complex it serves; poor transit service; and limited practical opportunity to walk or bike due to distance and safety. When we create a built environment for the automobile, a scale of development is required that makes it difficult for other modes of transportation such as biking, walking, and transit to be realistic alternatives.

16.4.2 Humans as Owners of Space

Owners face the complex problem of providing land use that the market desires, within the confines of what the government will allow and what is economically feasible to construct and operate. These three forces of market demand, economics of construction and operation, and government regulation frequently do not match up.

For land owners, organizations such as Leadership in Energy and Environmental Design (LEED) (Kibert 2012) and other best-practices organizations seek to help interested owners and developers produce a more sustainable built environment. Owners have direct control over what is built and how it is constructed within the limits of government regulations. Buildings consume a tremendous amount of energy to heat, cool, and light (besides other commercial or industrial processes that may be occurring within). Great strides have been made in building technologies to make both the construction and operation of buildings more sustainable—e.g., use less energy and produce less pollution. LEED and other best practice standards are rarely required by government, so their adoption and use is determined by the land owners/developers in consultation with the perceived market. The increased cost of construction is, at least hypothetically, more than compensated for by reduced energy costs and public good will in locations where energy and environmental sustainability are recognized issues. There are some instances where LEED and other best practices are either required or encouraged by government policy. Market preferences and return on investment gains will determine to what extent these practices and standards become more widely adopted.

16.4.3 Government

Cities and counties regulate land use through general or comprehensive plans, zoning ordinances, subdivision regulations, and a myriad of area specific or strategic plans. These plans and ordinances create the rules within which land development and land use occur in that jurisdiction. These rules can differ dramatically from one jurisdiction to another and ideally represent the preferences of the majority of citizens who live there. If land owners or users feel that these rules are unduly burdensome, arbitrary, capricious, or simply unconstitutional, they have the privilege of challenging them in court, seeking a finding and potential relief.

Private developers, chain stores, and other economic development compare these rules between cities to determine which work most in their favor (but within the constraints of the market) before selecting a site or location to develop. In general, if city X is perceived to have pro- growth laws compared to city Y, developers will select sites in city X for development. However, if the market conditions in city X are worse than city Y, they may agree to work with the more restrictive rules to access greater profits from the stronger market conditions.

Similarly, land users evaluate how local government taxes are spent, the quality of government provided services such as parks, schools, and infrastructure, and local tax structure before selecting locations to live and shop. If conditions are perceived to be "better" in city X than city Y they may relocate their residence or shift shopping locations to take advantage of the lower taxes or improved government services. This phenomenon is called "voting with feet"—a term and concept credited to Charles Tiebout (1956) simply stating the human actors are economically rational and have a choice of locations (cities, neighborhoods, etc.) and will choose to locate or move to locations that better meet their needs, desires, and preferences.

Often, advocates of sustainability champion strong government policy that requires additional sensitivity to the natural environment in the manner of construction, layout, and interconnection (roads and transit facilities) of our built environment. While these policies can do much good, they must be in alignment with local preferences and reflect local market realities in order to be effective.

16.4.4 Interactions Between Actors

Land users, land owners, and the government interact dynamically to make decisions regarding the built environment. Each comes to the table with a different perspective and set of powers they can use to influence the decisions of the others. All residents use land. As such, land users are the largest and most powerful group. However, individually, land users are the least powerful, having no direct control over the built environment on a significant scale.

As a group, land users represent the market to land owners and the voting body public to the government. In these roles, land users heavily influence the decisions of land owners and the government. Because power is derived from collective actions rather than individual ones, much effort is expended trying to form large enough groups of land users, or citizens, to exercise collective power. Political parties, neighborhood associations, religious and other charitable organizations, environmental interest groups, issue oriented campaign organizations, etc. all strive to get enough land users to collectively join their various efforts to bring about change.

The interactions between these groups can at times be contentious. Interests often conflict and finding mutually beneficial solutions can be challenging. With the manner of construction and operation largely the prerogative of the land users and owners, the location of differing land uses and the connections between land uses (i.e. the transportation system) falls predominately to the government. Geotechnologies such as GIS and remote sensing are used in the public, sustainability forum as tools for monitoring and forecasting the impacts of various land use, construction, and transportation decisions within the built environment. These tools form the information system that supports built environment decisions and debate. GIS and remote sensing also form the backbone of analysis tools such as travel demand models and integrated land use and transportation forecasting models. (More will be said about these analytical tools later in this chapter.)

16.4.5 City Planning and Sustainable Built Form

In the latter portions of the nineteenth century, urban areas were plagued with numerous human health issues. Among these were devastating fires, communicable diseases, and issues with waste removal. Cities were noisy, crowded, congested, polluted, and dangerous. The profession of City Planning came into being in the early part of the twentieth century with the mandate to fix these urban ills.

One of the first things city planners did was to separate incompatible land uses. This was largely done through a practice called "zoning" where the local government divided up the urban area into districts and assigned appropriate land uses to each district. In this way, noxious uses such as industrial or manufacturing could be kept away from residences and adult entertainment could be kept away from elementary schools. This type of built form, with rigid separation of land uses was a fairly new and novel approach to solving urban ills, and it worked well. While it required several decades for zoning to have a significant impact on the built environment in older cities, the impact could clearly be seen in rapidly growing cities of the mid-west and western portions of the United States.

The reason this rigid separation of land uses was not possible previous to this time was that for thousands of years prior to the early twentieth century, transportation consisted of walking, riding on an animal, or riding in a cart or carriage pulled by an animal. The speed of such transportation necessitated mixing urban land uses to keep all needed activities proximate to one another. Further, the lack of in-home refrigeration units meant that most families bought much of their food the same day they consumed it—with the exception of dry goods.

With the advent and then rapid adoption of technologies such as in-home refrigerators, freezers, and the private automobile, city planners were able to put great distances between noxious uses. The roads in this new built environment once used for walking and animal powered transportation gave way first to the electric street car and then to the private automobile. The invention of the intra-city truck enabled noxious land uses to be moved to the district zoned for such uses. The intercity truck enabled noxious land uses to be moved out of the city entirely. Global shipping enabled noxious land uses to be moved to different counties and even different continents. Under these conditions, the modern US city is far cleaner than its eighteenth and nineteenth century counterparts.

Urban systems are complex and often solving one set of problems give rise to a new set of unintended and unforeseen problems. So it was with traditional zoning and the rise of the private automobile. Traditional zoning, still widely practiced today, created such large distances between differing but complementary land uses that the options of walking, biking, or public transit became unrealistic in most suburban settings. Within this built form, residents not only needed a car to go somewhere, they needed a car to go anywhere or do nearly any thing. Getting a gallon of milk, for example, now required the use of a 5,000 lbs automobile burning non- renewable natural resources and producing toxic air pollution. Many of the noisy, smelly, congested conditions of the early city were replaced by the noise, smell, and congestion of the automobile.

Today, the United States has over 8.5 million lane miles of roads, 250 million private automobiles that drive a combined roughly 3 trillion miles per year. The average American household spends 10–20 % of its income on transportation related expenses and wastes on average 40 h a year (per person) stuck in congested traffic—in some cities that number is as high as nearly 70 h per year per person (Bureau of Transportation Statistics 2014).

For most of the twentieth century, traffic engineering as a discipline focused on getting the most vehicles across any given section of road as quickly as possible, with the goal being to maximize the vehicle "through-put" of the facility. Through-put is the most commonly used metric for evaluating the efficiency of a roadway. To maximize through-put, lanes needed to be wide, curves needed to be broad and sweeping, and opposing traffic needed to be placed on a separate facility. Crossing traffic needed to travel above or below on separate facilities that minimized or eliminated conflicts of movement between vehicles. From this approach the modern interstate highway was born, allowing vehicles to cross vast distances at great speeds. It was the summum bonum of through-put. Of course, pedestrians, bicycles, and other non-motorized forms of transportation could not be allowed on these facilities both for their own safety and for the sake of maximizing the efficiency of the system.

With an ever growing appetite for roads and driving and a limited amount of funds for road construction, a national standard for transportation planning was created (ITE 1999). The federally mandated transportation planning process is presented in Fig. 16.2. This standard did two things. First, it standardized practice and instantly became the curriculum in traffic engineering and transportation planning programs in universities across the country. Every graduating engineer/planner knew exactly how to plan an efficient transportation network that would maximize the through-put of automobiles. Second, it enabled road projects to be evaluated head-to-head to see which would provide the most benefit (though-put) per dollar spent. This model was designed to create 20 year forecasts of traffic conditions on the road network and is used in creating both short and long-range plans. It has been standard practice in transportation planning and transportation capital improvement selection since its inception in the 1960s.

There at least two major concerns with this model of transportation planning. First, the only set of questions it is designed to answer are "where should we build new road facilities?", "which existing roads or intersections should be improved or widened?", or "what improvements should be made to our mass transit system?" As such, the only available "answer" is a road improvement or transit investment somewhere in the system. To improve traffic network performance (the goal of the transportation planning process as it exists), a road needs to be built or widened, an intersection needs to be improved, or transit service needs to be upgraded through better system-wide coverage or shorter times between transit vehicles on a route (referred to as improvements in "headway"). Second, is that this model represents a static relationship between land use and transportation. This means that land use

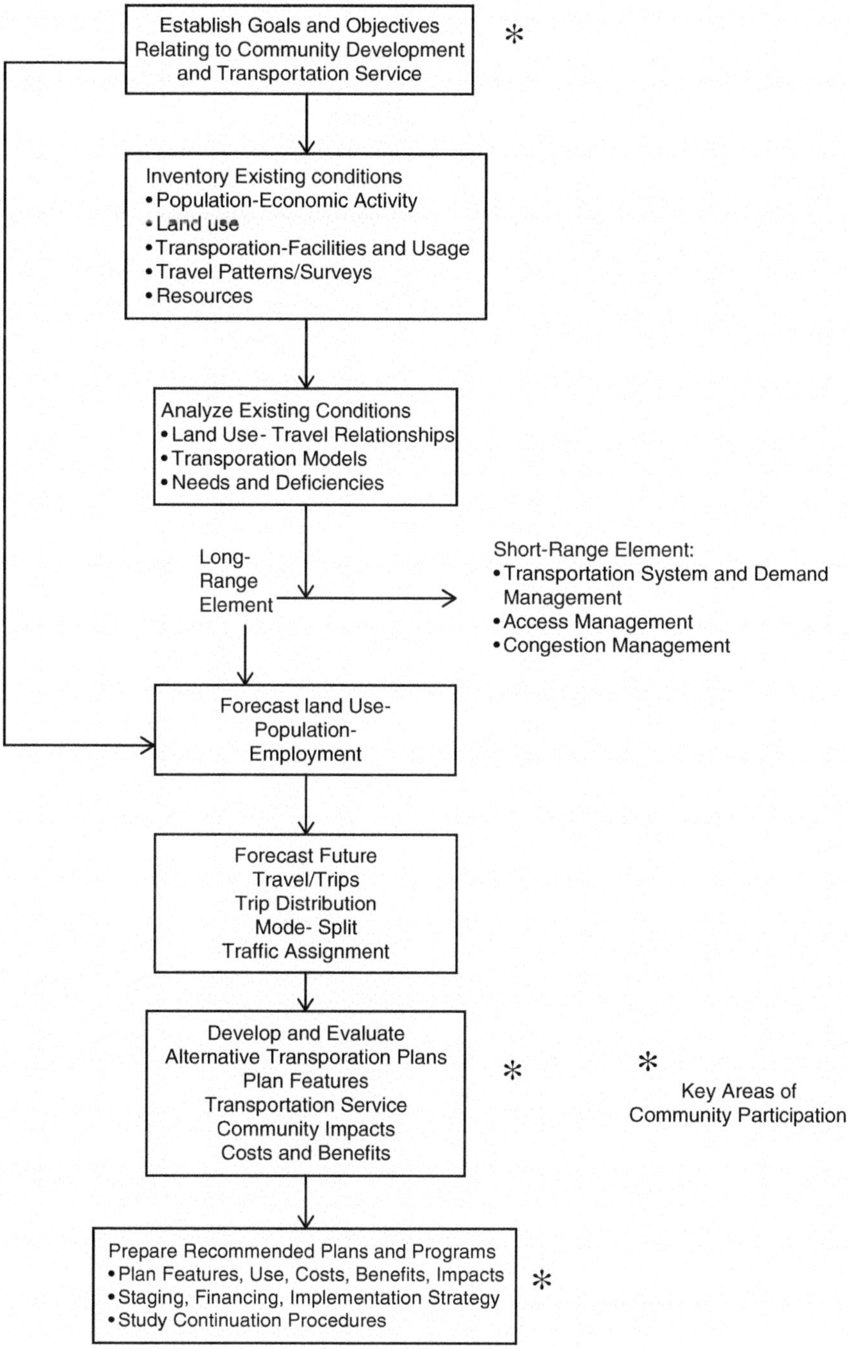

Fig. 16.2 Federally mandated transportation planning process (ITE 1999, pg. 97)

cannot shift or react in any way to changes in the transportation network or network performance. So, according the this model, you could build a new 12 lane interstate highway through the middle of a metropolitan area, vastly impacting performance of the travel network and land use would remain unchanged from the pre-construction condition. Of course, in reality land use does shift in response to road projects.

These two issues, combined with traditional zoning practice have created a problem commonly referred to as "sprawl". The natural consequence of the federally mandated transportation planning process is that planners evaluate existing conditions (including traffic conditions) and plan new roads or road improvements accordingly. Land owners respond to the free (to them), government provided improvements by shifting development paths to take advantage of the new or improved facilities. Land users occupy these new subdivisions and shop at the new retail space created along the new or improved roads. In doing so, they consume the excess capacity on the new roads and add further congestion to the network generally. The planner or engineer sees this new congested condition, runs it through his/her planning model and then plans new or expanded roads in an effort to alleviate this congestion. Because the model does not account for land use shifts, the new road projects appear to accomplish the congestion reducing goals for which they were planned and built. These new transportation facilities spur shifts in land use and the cycle repeats itself over and over again. During this process the planner/engineer feels confident they are making things better for the traveling public. This process created the massive suburban built form with rigid separations of land uses that we see today.

Sprawl is problematic from a sustainability standpoint for several reasons. First, the amount of land consumed per person has gone up dramatically. This is caused by increased wealth (which allows us to purchase more land, larger homes, and more cars), little to no growth controls in cities and suburbs (the nearly unlimited supply of land greatly reduces it market cost), massive government expenditures on roads (greatly decreasing the private costs of moving farther away from work and shopping locations), and the underpricing of travel and land. This last point warrants further explanation.

In market economics, all goods and services are assumed to bear a price that reflects their scarcity and the full market cost of the exchange. Externalities are inefficiencies that occur when the full market cost is not borne by the parties directly involved in the exchange. These inefficiencies cause us to over consume due to lower costs than the full market would demand. Our transportation system as it currently exists has at least three significant externalities: congestion, pollution, and accidents/vehicular deaths. The first several hundred vehicles per hour on a road do not negatively impact other cars on the road. Once a certain threshold is hit (which differs by road type—e.g. local road, collector, arterial, express way, highway, etc.) each new vehicle that enters the roadways causes other vehicles on the road to slow down. If vehicles continue to enter the roadway, capacity will be exceeded and the system will fail (i.e. stop and go traffic). The cars entering the system once it is near capacity place a time and convenience cost on all other drivers but do not compensate them for this cost. This is the congestion externality.

The pollution externality is caused when we consume fossil fuels in our cars. Roughly 13 million barrels of oil—a non-renewable natural resource—per day, creating roughly 1.9 billion metric tons of carbon dioxide every year from our vehicles (Bureau of Transportation Statistics 2014). The combustion process converts that fuel into energy, heat, and toxic exhaust. The energy is used to propel the car forward. The heat is taken care of by the vehicle's radiator and coolant system. The toxic exhaust is partially cleaned by the catalytic converter and the remainder is released into the air. These pollutants consists of two types, local air pollutants and global air pollutants. The local air pollution causes respiratory issues and illness. The global pollutants have been linked to climate change and will likely have effects we cannot now fully determine. The driver of the vehicle that created the pollution is not asked to pay the individuals harmed by his/her pollution. This is the pollution externality.

Every time an individual makes the decision to drive a motor vehicle, they have a chance of unintentionally using that vehicle to harm another person. While the probability is small for any individual, we still kill tens of thousands of individuals each year with our cars. We are not asked to pay each other for the chance we have of killing them each time we drive. This is the accident or death externality.

As mentioned above, the lack of growth controls on our urban areas causes land to be over supplied and therefore less expensive than it should be if it assumed the full market cost of owning and occupying it. For example, people moving outward, away from the urban core requires significant expenditures on constructing and maintaining infrastructure to serve them. There are very few examples in this country of single family detached homes that pay enough in property taxes to maintain the road infrastructure directly in front of them, from property line to property line. These costs are generally paid for by sales tax, which may be paid for by both residents and nonresidents of that municipality.

Our ever increasing amount of land consumed per person in our urban areas (through the twentieth century) has impacted the natural environment. As our cities sprawl outwards, the amount of natural habitat and animal migration corridors that are impacted increases. The full impact of this on human quality of life cannot be calculated.

The challenge of addressing these externalities is: (1) calculating what the market costs are on an individual basis; (2) determining a realistic way of charging individuals for these externalities; and (3) figuring out a mechanism of compensating individuals for the externalities they experience from the actions of others. The point of bringing these externalities up isn't to solve any of these issues but rather to illustrate that we are not paying the full market cost of driving and are therefore driving too much—any time a good is underpriced, it will be over consumed.

16.5 Approaching Decision-Based Solutions

As the humans whose choices shape our built environment, we can choose a more sustainable urban future. This section contains a few recommendations for tools, policies, and approaches that can improve the sustainability of our built environment.

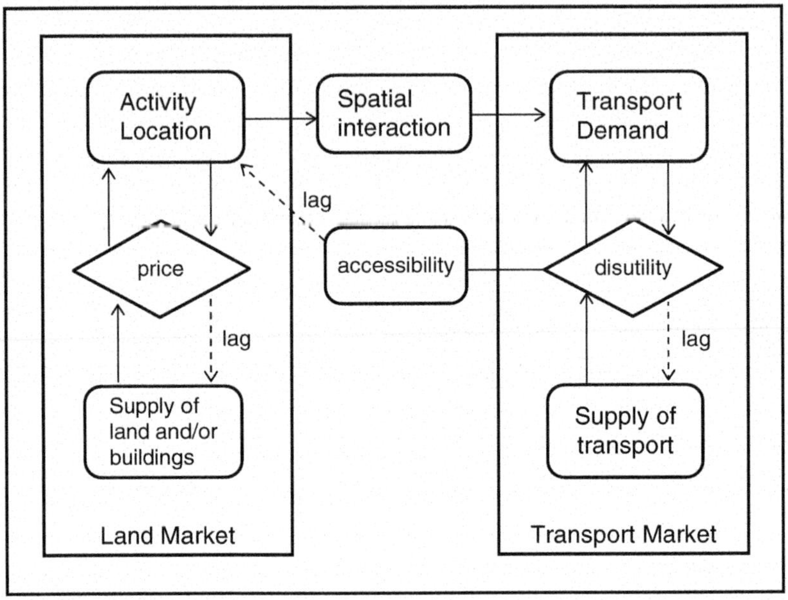

Fig. 16.3 Integrated land use and transportation planning model (Reproduced with permission from Abraham 2000)

16.5.1 Integrate Land Use Dynamically with Transportation Planning Models

Conceptually, the simplest way to "fix" deficiencies in the current transportation planning model is to give land use a dynamic relationship with transportation—allowing each to impact the other. Currently, land use is an input into the transportation planning process that is not affected by the various road construction projects that the model is asked to evaluate. Obviously the accessibility or improved traffic conditions new roads provide impacts where various land uses want to locate. The shift in land use that results from the changes in accessibility then affect the traffic patterns on the roads. This cyclical relationship is typically represented in integrated models as the interaction of two markets: the market for land and the market for transportation (see Fig. 16.3). In this model, price determines what land uses go where and travel times determine which mode (e.g. walk, drive, transit, etc.) of transportation and which route (on the street network or transit route) on that mode we take. Once a single model year has arrived at a stable market equilibrium, the transportation accessibility data are forwarded to the next model year as input information for the land use market (see Fig. 16.4). The model repeats this process year by year until it arrives at the planning "horizon" year (typically 20 years into the future). Transportation projects can then be programmed into the modeling process at the appropriate year and the land use and transportation systems can respond to the changes in the road network with shifts in land use and shifts in

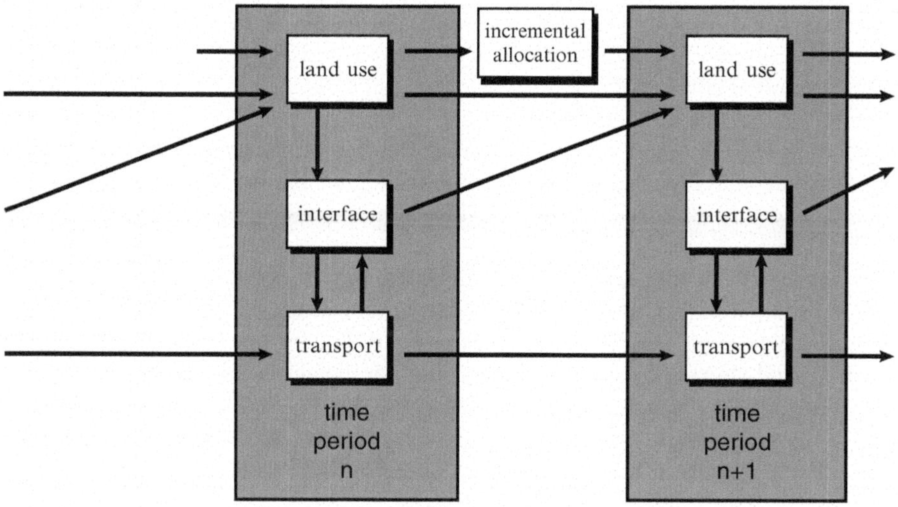

Fig. 16.4 Temporal relationships and data flow in the integrated land use and transportation planning model (Reproduced with permission from Abraham 2000)

travel patterns as applicable. Planning models that integrate land use and transportation dynamically have been shown to produce superior results when compared with the traditional approach (Hunt et al. 2001; Rodier 2009; Conder and Lawton 2002). By including land use dynamically, the planning model is also able to test a broader variety of policies, including land use policies as alternative solutions to new or expanded roads (Weidner et al. 2005; Johnston et al. 2005; Abraham and Hunt 1999).

16.5.2 Form-Base Zoning

Traditional zoning policies contribute to sprawl by rigidly separating land uses and creating distances that can only be traversed with motorized forms of transportation. This produces excess consumption of non-renewable natural resources and increased air pollution. Initially there was wisdom in separating noxious uses from residential uses; it increased quality of life. Today, many noxious uses have been completely eliminated from our urban areas. Those noxious uses that remain should still be segregated. This is largely done by creating industrial parks or other specific areas of the city where heavy industrial or manufacturing activities are performed. Most remaining uses, however, could likely tolerate each other in appropriate proportions. By mixing retail, office, institutional (e.g. government and education), and residential land uses together, our built form could be more friendly to alternative modes of transportation such as walking, biking, and public transit.

Mixing land uses alone is not sufficient to bring about a sustainable, walkable built environment. The design of the city itself would have to change. Currently, in most sub-urban environments the amount of land dedicated to roads, intersections, parking lots, interstate highways, and interchanges far exceeds the amount of productive space (buildings). This built form, even in a mixed land use condition would not facilitate walking, biking, and public transit as viable modes of transportation. The distances between land uses would remain inaccessible to all but the private automobile.

There is a growing movement of land use/zoning reform called New Urbanism (Duany et al. 2000; Leinberger 2009). New Urbanism uses a form-based approach to zoning where the design of the buildings and their position with respect to the sidewalk and street are what is regulated and not the activities that go on inside of them. The idea is that having a professional office (e.g. medical, accounting, legal, etc.), school, or small retail establishment next to a residence would likely not be harmful to the residence. It would allow the residence to walk to these type of land uses as they would now be spatially proximate.

New Urbanism has also embraced the ideas behind the Complete Streets movement (McCann and Rynne 2010). Compete Streets seeks to make our roadways more balanced. Currently, roads are essentially built for cars and all other forms of transportation are secondary, at best. Complete Streets desires to have street plans include larger sidewalks for pedestrians, bicycle lanes, dedicated transit facilities, and travel lanes for cars. By combining New Urbanism's form- based zoning code with complete streets, we could create a more pedestrian friendly and sustainable built environment. Similar to integrated land use and transportation planning models, these would need to be policies adopted at the local level and supported by regional transportation planning agencies to take effect. These agencies are responsible to their citizens for the decisions they make.

16.5.3 Citizen Participation

In the United States we are fortunate to live in a representative democracy in which our voices can greatly influence the governmental agencies responsible for our built environment. Sadly, few people choose to participate in this process. People lead busy lives and typically, city council meetings, planning commission meetings, zoning board of adjustment meetings, regional transportation planning meetings, and other meetings held to plan for the built environment lack citizen participation. Legally, all agencies charged with land use or transportation planning duties are required to advertise their meetings, open their meetings to the public, and hold public hearings in which citizens are allowed to participate.

A representative democracy is greatly benefited by broad citizen participation in which the views of the people, as divergent as they may be, are voiced. Representative democracy is further benefited by citizens who take the time and effort to become informed of the issues and what the long term impacts of different approaches to solving these issues will likely be. Representative democracy is benefited by citizens who vote. Voter turnout in most local elections is quite small. This means that the people ultimately making decisions regarding the built environment can be elected by a relatively small portion of the population, leaving local elections susceptible to special rather than common interests.

The sustainability of our built environment is largely in the hands of local citizens (both as land users and land owners) and local government. The federal government occasionally requires auto manufactures to build more fuel efficient cars. This has only happened twice in our nation's history. Auto manufactures complain that if consumers wanted more fuel efficient cars, the government would not have to mandate it. Further, cleaner fuels are generally a state-level policy choice. Other than these two instances, local governments, led by their citizens, choose how sustainable their built environment will be.

16.6 Conclusion

Our built environment is perhaps our biggest piece of collective, public art that we will ever create. It represents us more honestly than we could represent ourselves. It is as if all of our preferences, desires, hopes, dreams, and the best and worst of our natures was neatly placed upon a canvas that we call a city. Our cities are formed by us, owned by some of us, and governed by all of us. By-in-large we have done a good job. Our built environment has improved our quality of life, access to education, healthcare, employment, learning, and social opportunities. Our cities have gotten cleaner, healthier, and safer over time.

There remains much room for improvement in our built environment, particularly on the sustainability front as has been laid out in this chapter. That work is ours to undertake lest future generations be left with diminished opportunities due to decreased reserves of energy rich natural resources. There is much we can do to improve the quality and sustainability of our cities. This chapter has presented a few modest measures that individuals and communities could take to improve local government decision making and the long-term sustainability of our built environment. Urban areas are built by human choices. Human choices will direct the future of our cities for ourselves and our posterity.

References

Abraham J (2000) Parameter estimation in urban models: theory and application to a land use transport interaction model of the Sacramento, California Region. Diss. University of Calgary, Department of Civil Engineering. http://hbaspecto.com

Abraham J, Hunt JD (1999) Policy analysis using the Sacramento MEPLAN land use transportation interaction model. Transp Res Rec J Transp Res Board 1685:199–208

Bradford HD (2009) Blueprint for disaster: the unraveling of Chicago's public housing. University of Chicago Press, Chicago

Bureau of Transportation Statistics (2014) Pocket guide to transportation 2014. US Department of Transportation, Washington, DC

Conder S, Lawton K (2002) Alternative futures for integrated transportation and land use models contrasted with Trend-Delphi models: Portland Oregon metro results. Transp Res Rec J Transp Res Board 1805:99–14

Duany A, Plater-Zyberk E, Speck J (2000) Suburban nation: the rise of sprawl and the decline of the American dream. North Point Press, New York. Print

Hunt JD, Johnston R, Abraham JE, Rodier CJ, Garry GR, Putman SH, de la Barra T (2001) Comparisons from Sacramento model test bed. Transp Res Rec J Transp Res Board 1780:53–63

Institute for Transportation Engineers (1999) Transportation planning handbook, 2nd edn. Institute for Transportation Engineers, Washington, DC

Johnston RA, Gao S, Clay MJ (2005) Modeling long-range transportation and land use scenarios for the Sacramento region, using citizen-generated policies. Transp Res Rec J Transp Res Board 1902:99–106

Kibert C (2012) Sustainable construction: green building design and deliver, 3rd edn. Wiley, Hoboken

Leinberger CB (2009) The option of urbanism: investing in a new American dream. Island Press, Washington, DC

McCann B, Rynne S (2010) Complete streets: best policy and implementation practices. American Planning Association, Chicago

O'Sullivan A (2006) Urban economics, 6th edn. McGraw-Hill/Irwin, New York

Rodier C (2009) Review of International Modeling Literature: transit, land use, and auto pricing strategies to reduce vehicle miles traveled and greenhouse gas emissions. Transp Res Rec J Transp Res Board 2132:1–12

State v. Diamond Motors, Inc. (1967) 429 P.2d 825

Tiebout CM (1956) A pure theory of local expenditures. J Polit Econ 64:416

Vale LJ (2013) Purging the poorest: public housing and the design-politics of twice cleared communities. University of Chicago Press, Chicago

Village of Euclid (1926) Ohio v. Ambler Realty Co., 272 U.S. 365

Weidner T, Gregor B, Wert M, Hunt JD (2005) Oregon bridge investment alternatives: using integrated modeling in policy decisions. Presented at the 84th Annual Meeting of the Transportation Research Board, Washington, DC

Chapter 17
Geotechnologies & Sustainable Urban Environments

Ryan R. Jensen

17.1 Volume Objective

The objective of this volume was to focus on and collect a range of real world case studies, novel examples, and innovative approaches towards urban sustainability – particularly focusing on research using geotechnologies and/or a geographic framework. Ideally, the volume will be used by academic and professional researchers, community leaders, planners, students, and anyone else who has a desire to better assess urban environmental change. Additionally, the volume examines the prospect of deploying geotechnologies, broadly conceptualized, to understand and enhance the sustainability of urban systems. The chapters included examples of cultural/social adaptation, political-ecology, the impact of little investigated industries, economic development, land use change, and human-environment interactions. In contrast to other books on the subject, which tend to focus on planning, engineering, environmental impacts, or political discourses (i.e., political ecology), this volume unites the three pillars of sustainability – socio-political/cultural, economic, and environmental change – by focusing on spatial methodologies and emerging technologies.

While the studies vary widely, the collection has provided examples from each domain. Some chapters described specific research methods and projects (e.g., Lawrence; Fuller; Jensen and Hardin; Chohaney et al.; Hagelman et al.; Lascell; McLean et al.). Other chapters described concepts and practices more generally (e.g., Reid; Lulla and Jensen; Gallagher and Lawrence; Clay and Albers) even while discussing these things or strategies in specific areas or industries (e.g., Metcalf et al.; Patterson et al.; Knigge et al.). In many respects, the depth and breadth of these studies is impressive and underscore the complexity of the urban environment

R.R. Jensen (✉)
Department of Geography, Brigham Young University, Provo, UT, USA
e-mail: ryan.jensen@byu.edu

J.D. Gatrell et al. (eds.), *Urban Sustainability: Policy and Praxis*,
Geotechnologies and the Environment 14, DOI 10.1007/978-3-319-26218-5_17

and the many different ways that geotechnologies can be implemented to study and disentangle the complexities of sustainability in place.

17.2 Future Research

Insofar as urban change will remain a significant topic in both public and academic dialogues for many years to come, the practice of using geotechnologies to study, explain, and maintain sustainable urban environments will continue. As such, geotechnologies have much to offer researchers, planners, government officials, and many others as they seek to understand these very complex processes and the intersections between social and physical processes. Specifically, future research could focus on using geotechnologies in a variety of urban environments to study ecosystem characteristics and services and their relationship to urban sustainability. These studies might be completed in both large and small urban areas with differing socio-demographic characteristics across multiple ecosystems. Further, comparative studies that examine different urban areas (e.g., high and low population density) and the value and character of urban ecosystem services in these different spaces and their relationships to sustainability would be enlightening.

Finally, we, the co-editors, recognize that any time and effort used to explore and understand the urban environment is time and effort well spent. As humans continue to migrate to cities, it is useful to examine the urban environment and the properties of individual communities and neighborhoods – especially as these things relate to sustainable urban environments (Jensen et al. 2004). As noted in the introductory chapter, this book is not exhaustive in its presentation of ways to use geotechnologies to study the urban environment and urban systems. However, after reading the chapters in this book, I earnestly believe that readers will find new and interesting ways to study the urban environment. More importantly, I believe the chapters provide new insights into sustainability at the micro to the macro scales across urban communities, the industries that support growth, and the physical geographies within which cities continue to evolve and transform. In short, I and the co-editors expect that future researchers will continue to use geotechnologies, specifically, and spatial approaches, generally, to study humanity's principal habitat.

Reference

Jensen R, Gatrell J, Boulton J, Harper B (2004) Using remote sensing and geographic information systems to study urban quality of life and urban forest amenities. Ecol Soc 9(5):5. Online: http://www.ecologyandsociety.org/vol9/iss5/art5/; Last accessed 8 June 2015

The manufacturer's authorised representative in the EU is Springer Nature Customer Service Centre GmbH, Europaplatz 3, 69115 Heidelberg, Germany. If you have any concerns regarding our products, please contact ProductSafety@springernature.com

Printed and bound by CPI Group (UK) Ltd, Croydon, CR0 4YY
05/10/2026
02228134-0002